Antibiotic Resistance

Antibiotic Resistance

Understanding and Responding to an Emerging Crisis

Karl Drlica
David S. Perlin

Vice President, Publisher: Tim Moore
Associate Publisher and Director of Marketing: Amy Neidlinger
Acquisitions Editor: Kirk Jensen
Editorial Assistant: Pamela Boland
Operations Manager: Gina Kanouse
Senior Marketing Manager: Julie Phifer
Publicity Manager: Laura Czaja
Assistant Marketing Manager: Megan Colvin
Cover Designer: Sheila Schroeder
Managing Editor: Kristy Hart
Project Editor: Anne Goebel
Copy Editor: Apostrophe Editing Services, Inc.
Proofreader: Sheri Cain
Indexer: Erika Millen
Senior Compositor: Gloria Schurick
Manufacturing Buyer: Dan Uhrig

Publishing as FT Press
Upper Saddle River, New Jersey 07458

FT Press offers excellent discounts on this book when ordered in quantity for bulk purchases or special sales. For more information, please contact U.S. Corporate and Government Sales, 1-800-382-3419, corpsales@pearsontechgroup.com. For sales outside the U.S., please contact International Sales at international@pearson.com.

Printed in the United States of America

First Printing January 2011

ISBN-10: 0-13-138773-1
ISBN-13: 978-0-13-138773-7

Pearson Education LTD.
Pearson Education Australia PTY, Limited
Pearson Education Singapore, Pte. Ltd.
Pearson Education Asia, Ltd.
Pearson Education Canada, Ltd.
Pearson Educación de Mexico, S.A. de C.V.
Pearson Education—Japan
Pearson Education Malaysia, Pte. Ltd.

Library of Congress Cataloging-in-Publication Data

Drlica, Karl.
Antibiotic resistance : understanding and responding to an emerging crisis / Karl Drlica, David S. Perlin.
p. ; cm.
Includes bibliographical references and index.
ISBN-13: 978-0-13-138773-7 (hardcover : alk. paper)
ISBN-10: 0-13-138773-1 (hardcover : alk. paper) 1. Drug resistance in microorganisms.
I. Perlin, David. II. Title.
[DNLM: 1. Drug Resistance, Microbial—immunology. 2. Immunity, Innate.
QW 45 D782a 2011]
QR177.D75 2011
616.9'041—dc22
2010014841

We thank our families for their support and dedicate this work to the patients and clinicians who are confronting the harsh reality of drug-resistant infections.

Contents

About the Authors

Karl Drlica, Ph.D. is a Principal Investigator at the Public Health Research Institute and Professor of Microbiology & Molecular Genetics at the UMDNJ—New Jersey Medical School in Newark, New Jersey. Dr. Drlica's laboratory focuses on fluoroquinolone action and resistance with *Mycobacteriun tuberculosis* and other bacteria, including approaches for slowing the enrichment and amplification of resistant bacterial subpopulations.

David S. Perlin, Ph.D. is Executive Director of the Public Health Research Institute and UMDNJ Regional Biocontainment Laboratory, as well as Professor of Microbiology & Molecular Genetics at the New Jersey Medical School in Newark, New Jersey. He is also a Fellow of the New York Academy of Sciences. Dr. Perlin's laboratory explores mechanisms of antifungal drug resistance, rapid detection of drug-resistant bloodstream pathogens in high-risk patients, and the application of small-animal models for the study of respiratory pathogens.

Preface

Recent human activities have profoundly influenced our global environment, often in ways we did not anticipate. An example is our use of antibiotics. Initially hailed as "magic bullets," these chemical agents are now used so often that success threatens their long-term utility. Unfortunately, the natural mutability of microbes enables pathogens to develop bulletproof shields that make antibiotic treatments increasingly ineffective. Our failure to adequately address resistance problems may ultimately push the control of infectious disease back to the pre-penicillin era. Indeed, it is now impractical to simply invent additional antibiotics to replace those lost to resistance. However, ideas have emerged for slowing the development of antibiotic resistance in individual patients and in the human population as a whole. *Antibiotic Resistance* introduces these ideas.

Antibiotic Resistance was initially drafted to supplement studies of infectious disease. The problem of resistance tends to be neglected, which puts the well-being of our society at increasing peril. In the course of completing this book, we realized that everyone makes decisions about antibiotic use; therefore, everyone needs to understand how human activities contribute to resistance. Individual patients, medical providers, and agricultural specialists all have a role to play in providing a safer environment. We now aim to make the principles of antibiotic use and effectiveness available to a large audience: farmers, hospital administrators, government regulators, health department personnel, pharmaceutical executives, and especially individual users. (Individual patients pressure their doctors for treatments, and in most cases, patients decide whether to take medicines as prescribed; in countries where prescriptions are not required to purchase antibiotics, patients are major decision makers.) Such diversity in readership poses a challenge.

Fortunately, detailed descriptions of chemical structures, molecular mechanisms, and epidemiological modeling are not required to understand the principles of resistance. We focus on broad concepts supported by examples and descriptions of key experiments. We expect that *Antibiotic Resistance* will be a quick read for persons with knowledge of biology. Those readers can then build on the principles with follow-up reading. Lay readers may find that some terms need to be defined. For them, we have provided a glossary and appendixes covering background concepts.

Our goal with *Antibiotic Resistance* is to point out how human activities contribute to the problem of resistance. Our hope is that an understanding of the complex factors involved in resistance will lead to changes that lengthen antibiotic life spans. An example of the complexity is seen in the traditional practice of setting antibiotic doses only high enough to cure disease. We argue that this practice encourages the emergence of resistance, that more stringent antibiotic regimens are needed to preempt the emergence of resistance. But from an individual patient perspective, using higher doses seems excessive when milder treatment usually cures disease. Why should the individual patient risk toxic side effects to preserve antibiotics for the general population?

Antibiotic waste disposal problems are also complex. In principle, environmental contamination with antibiotics exerts selective pressure on microbes. That pressure can lead to the evolution of resistance genes that then spread from one organism to another and eventually reach human pathogens. We do not know how often this scenario occurs, whether it is reversible, or how much we need to improve agricultural and hospital disposal programs to stop the process.

Fortunately, many resistance issues are not complex. For example, wearing contaminated gloves can spread drug-resistant disease in hospitals: More attention to hand hygiene is required. We are confident that an improved understanding of antibiotic resistance can help preserve these valuable agents.

Each year, thousands of scientific papers are published on antibiotic resistance, making it difficult for even a pair of authors to get everything right. To improve accuracy, we obtained help from David Alland, Vivian Bellofatto, Arnold Bendich, Purnima Bhanot, John Bradley, Dorothy Fallows, Alexander Firsov, Patrick Fitzgerald, Marila Gennaro, Tao Hong, Dairmaid Hughes, Robert Kerns, Barry Kreiswirth, Shajo Kunnath, David Lukac, Simon Lynch, Muhammad Malik, Barun Mathema, Ellen Murphy, Christina Ohnsman, Richard Pine, Lynn Ripley, Snezna Rogelj, Bo Shopsin, Ilene Wagner, Heinz-Georg Wetstein, Xilin Zhao, and Stephen Zinner. We sincerely thank them for their time and for sharing their knowledge.

Chapter 1

Introduction to the Resistance Problem

Summary: As a normal part of life, we are all exposed to pathogens, the tiny microbes and viruses that cause infectious disease. Many pathogen varieties exist. Some are even harmless inhabitants of our bodies most of the time. A common feature of pathogens is their microscopic size. Another is the huge numbers their populations can reach during infection, often in the millions and billions. Human bodies have natural defense systems, but those systems sometimes fail to control infection. For such occasions, pharmaceutical companies have developed antibiotics, chemicals that interfere with specific life processes of pathogens. As a natural response, antibiotic resistance emerges in pathogen populations. Resistance is a condition in which the antibiotic fails to harm the pathogen enough to cure disease. Emergence of resistance often begins with a large pathogen population in which a tiny fraction is naturally resistant to the antibiotic, either through spontaneous changes or through the acquisition of resistance genes from other microbes. Antibiotic treatment kills or halts the growth of the major, susceptible portion of the microbial population. That favors growth of resistant mutants. Prolonged, repeated use of a particular antibiotic leads to the bulk of the pathogen population being composed of resistant cells. Subsequent treatment with that antibiotic does little good. If the resistant organisms spread to other persons, the resulting infections are resistant before treatment: Control of such infection requires a different antibiotic. The development of resistance is accelerated by the mutagenic action of some antibiotics, by the movement of resistance genes from one microbial species to another, and by our excessive, inappropriate use of antibiotics. In the past, a successful medical strategy was to develop new, more potent antibiotics. However, the pharmaceutical pipeline to new antibiotics is no longer adequate.

In this chapter, we define terms and provide an overview of antibiotic resistance. One of the key problems is that as a global community we have not considered antibiotics as a resource to be actively protected.[1] Consequently, we use antibiotics in ways that directly lead to resistance. Changing those ways requires an understanding of antibiotic principles. We begin with a brief description of MRSA to illustrate a bacterial-based health problem.

MRSA Is Putting Resistance in the News

MRSA is the acronym for methicillin-resistant *Staphylococcus aureus*. (Acronyms are usually pronounced letter by letter, as in DNA; scientific names are always italicized; after an initial spelling of the entire name, the first name is often abbreviated by its first letter.) *S. aureus* is a small, sphere-shaped bacterium (see Figure 1-1) that causes skin boils, life-threatening pneumonia, and almost untreatable bone infections. It often spreads by skin-to-skin contact, shared personal items, and shared surfaces, such as locker-room benches. When the microbe encounters a break in the skin, it grows and releases toxins.

Figure 1-1 *Staphylococcus aureus.* Scanning electron micrograph of many MRSA cells at a magnification of 9,560 times.

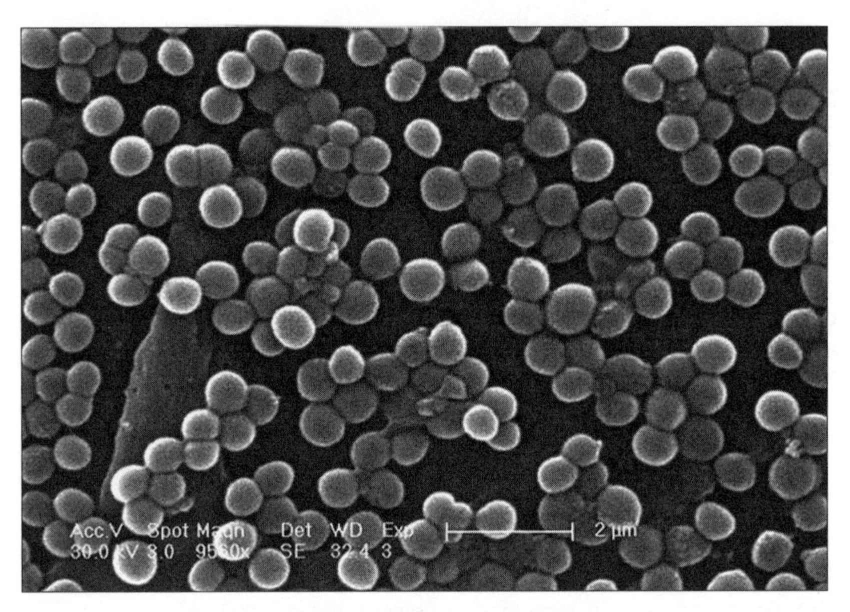

Public Health Image Library # 7821; photo credit, Janice Haney Carr.

Sixty years ago, *S. aureus* was very susceptible to many antibiotics, including penicillin. Susceptibility disappeared, and the pharmaceutical industry produced increasingly potent antibiotic derivatives. Among these was methicillin, which overcame resistance to penicillin. But in 1960, one year after the introduction of methicillin, MRSA was recovered in the United States. As the resistant bacterium spread through hospitals, surgical procedures and long-term use of catheters became more dangerous. MRSA also caused pneumonia, commonly following influenza, and recently skin infections caused by MRSA captured public attention. In one newspaper account,[2] pimples on a newborn baby were found to contain MRSA. Antibiotics cleared the infection; however, a month later, the father found boils on his own leg that contained MRSA. Treatment cleared the boils, but they came back. The mother developed mastitis during breast feeding that required a 2-inch incision into her breast to drain the infection. About a year later, an older child developed an MRSA boil on his back. The family is now constantly on alert for MRSA, trying to wash off the bacteria before the microbes find a break in the skin.

Community-associated MRSA has its own acronym (CA-MRSA) to distinguish it from the hospital-associated form (HA-MRSA). Many community-associated *S. aureus* strains are members of a group called USA300, which now accounts for half of the CA-MRSA infections. The strain causes necrotizing (flesh-eating) skin infection, pneumonia, and muscle infection. In 2005, MRSA accounted for more than 7 million cases of skin and soft tissue infection seen in outpatient departments of U.S. hospitals.[3] As expected, CA-MRSA strains are moving into hospitals. In a survey of U.S. hospitals taken from 1999 through 2006, the fraction of *S. aureus* that was resistant to methicillin increased 90%, almost entirely from an influx of CA-MRSA.[4]

Although many infections tend to occur in persons having weakened immune systems, MRSA can infect anyone. For example, healthy young adults tend to be susceptible to a lethal combination of influenza and MRSA pneumonia. In Chapter 7, "Transmission of Resistant Disease," we describe occurrences of CA-MRSA infection among athletes. Fortunately, most of these dangerous CA-MRSA strains are still susceptible to several antibiotics; however, that susceptibility may soon disappear.

HA-MRSA has been a problem in hospitals for years; in many countries, it is getting worse. For example, in the United States, MRSA climbed from 22% of the *S. aureus* infections in 1995 to 63% in 2007 (from 1999 through 2005, it increased 14% per year).[5] From 2000 to 2005, MRSA helped double the number of antibiotic-resistant infections in U.S. hospitals, which reached almost a million per year or 2.5% of hospitalizations.[6] In the United States, more persons now die each year from MRSA (17,000) than from AIDS.

MRSA in hospitals is largely an infection-control problem, that is, control requires keeping the organism from spreading from one patient to another, and if possible, keeping it out of the hospital entirely. Neither is easy. For many years, the Dutch have had an aggressive screening program for incoming patients. They isolate persons who test positive for MRSA and treat them with antibiotics that still work with *S. aureus*. Entire wards of hospitals are closed for cleaning when an MRSA case is found, and colonized healthcare workers are sent home on paid leave until they are cleared of the bacterium. The cost is about half that required to treat MRSA blood-stream infections;[7] consequently, the effort is thought to be cost-effective.

Until recently, many U.S. hospitals took a different approach: MRSA infections were considered part of the cost of doing business. Holland is a small country that can implement specialized care—the United States has a much higher incidence of MRSA. Nevertheless, in 2007, a Pittsburgh hospital reported that it had adopted the Dutch method. The hospital saved almost $1 million per year by screening patients and by insisting on more intensive hand-washing protocols for hospital staff.[8] Other U.S. hospitals are reconsidering their own stance.

Individual consumers will begin to search for hospitals having low MRSA incidence. That search will be easier when hospitals publish their drug-resistant infection statistics. Some states now require reporting of MRSA to health departments; consequently, the numbers are being collected. As an added incentive for MRSA control, some insurance carriers refuse to cover hospital costs when a patient contracts MRSA while there. Hospitals have responded by setting up antibiotic oversight committees to help keep resistance under control.

Humans Live with Many Pathogens

MRSA is one type of pathogen, the collective word applied to microbes and viruses that cause disease. (The term microbe includes bacteria, some types of fungi, and protozoans.) Each type of microbe has a distinct lifestyle. Bacteria are single-celled organisms that reproduce by binary fission; each cell grows and then divides to form two new cells. Bacteria cause many of the diseases that make headlines: tuberculosis, flesh-eating disease, and anthrax. Pathogenic fungi include yeasts and molds. Yeasts are single-celled, whereas molds tend to grow as thread-like structures composed of many cells. (Some pathogenic fungi switch between the forms in response to the environment.) Yeasts and molds cause pneumonia, and in immuno-suppressed persons yeasts and molds can cause deadly systemic infections. Pathogenic protozoans, such as the types that cause malaria, are single-celled microbes that are often spread by insect bites. In tropical and subtropical regions, protozoan diseases are among the major killers of humans. Protozoa and helminths (worms) are usually called parasites rather than pathogens due to their larger size. In *Antibiotic Resistance*, we do not distinguish between pathogens and parasites, because antibiotics are used for maladies caused by parasites as well as by pathogens.

Viruses differ qualitatively from the cellular organisms just mentioned. Viruses cannot reproduce outside a host cell. They require the machinery of a living cell to make new parts. Indeed, one could argue that viruses are not alive even though they are composed of the same types of molecules found in microbes, plants, and animals. Another feature of viruses is that they are generally much smaller than microbes: An electron microscope is required to see most virus particles, whereas a light microscope is adequate for microbes.

Many microbes and viruses are found in and on our bodies (see Box 1-1). Some are beneficial; others are harmful. Some pathogens only occasionally cause infectious symptoms. For example, *Mycobacterium tuberculosis* enters a dormant state in most persons it infects, with a minority of infected persons exhibiting symptoms. However, immune deficiency enables *M. tuberculosis* to exit dormancy and cause disease. Other serious diseases arise from microbes, such as the yeast *Candida albicans*, that ordinarily live harmlessly in or on humans. This organism causes vaginitis with healthy women and more serious disease with immune-compromised patients.

Pathogens that normally grow only inside humans often have effective means of transmission. *Mycobacterium tuberculosis* and influenza virus are two that spread through air; *Vibrio cholerae*, the cause of cholera, contaminates drinking water; and many digestive tract pathogens move with contaminated food. (*Salmonella typhi*, the bacterium that causes typhoid fever, is an example.) Many other pathogens are spread by insects and ticks. Among these are the protozoans responsible for sleeping sickness and malaria, the bacteria that cause plague and typhus, and many types of viruses, such as the agent of yellow fever. Avoiding contact with pathogens is exceedingly difficult.

Box 1-1: Pathogen Diversity

The scientific literature lists about 1,400 species of human pathogen: 538 bacteria, 317 fungi, 287 helminths, 208 viruses, and 57 protozoa. Over the last 20 years, almost 180 species either increased their incidence in humans or are expected to do so shortly. Only a small number, probably fewer than 100, cause disease *only* in humans. Almost 60% of human pathogens are zoonotic, that is, they move between humans and other vertebrates. Most of the others are commensals that usually live in or on humans without harm or are environmental organisms, living in water or soil. As we change our behavior and environment, new diseases emerge, largely through a species-jump from animal to human. Because human societies continue to evolve and change their interactions with animals, we are continually faced with new infectious diseases. For example, changes in food production led to the mad cow disease problem, the exotic pet trade led to monkeypox outbreaks, and harvesting bush meat (monkeys, and so on) probably led to infection with a virus that evolved into human immunodeficiency virus (HIV).[9,10]

Antibiotics Block Growth and Kill Pathogens

Antibiotics are drugs, taken orally, intermuscularly, or intravenously, that counter an infection. They include agents such as penicillin, tetracycline, ciprofloxacin, and erythromycin. Common bacterial diseases treated with antibiotics are tuberculosis and gonorrhea. Fungal and protozoan diseases are also treatable, but with agents specific for these organisms. (The biochemistry of fungi and protozoa differs substantially from that of bacterial cells.) Antiviral agents constitute a third set of specialized compounds. In general, little cross-reactivity exists among the categories, that is, agents used for fungi do not cure infections caused by viruses, bacteria, or protozoa. However, the principles underlying action and resistance are the same; consequently, in *Antibiotic Resistance* we lump all these agents together as antibiotics. Combining all the agents into a single category risks confusion, because the public has been told repeatedly not to use antibiotics for viral diseases. In this instruction, antibiotics are equated to antibacterials, and indeed antibacterials should not be used for viral infections. But the world is changing. We now have many antiviral and antifungal agents that are just as antibiotic as penicillin. The important issue is to identify principles that enable experimental data obtained with one agent to be used for making decisions with another. Such a cross-disciplinary effort is facilitated by having a general term (antibiotic); we use specific terms, such as antibacterial and antiviral, only when we need to distinguish the agents.

In molecular terms, antibiotics are small molecules that interfere with specific life processes of pathogens. Antibiotics generally enter a pathogen, bind to a specific component, and prevent the component from functioning. In cases of lethal antibacterials, treatment leads to formation of toxic reactive oxygen species that contribute to bacterial death. Not all antibiotics kill pathogens. Indeed, many of the older drugs only stop pathogen growth. Nevertheless, they can be quite effective because they give our natural defense systems time to remove the pathogens.

Antibiotics have been called magic bullets and miracle drugs because they quickly cure diseases that might otherwise cause death. When penicillin first became available in the middle of World War II, it gave life to soldiers who were otherwise doomed by infection of minor wounds. Penicillin was so valuable that urine was collected from treated soldiers and processed to recover the drug. Now antibiotics enable many complicated surgeries to be performed without fear of infection. Developments in molecular biology have even enabled pharmaceutical companies to design antibiotics that work against viruses. Among the more striking examples are antibiotics that attack the human immunodeficiency virus (HIV): They reduce the viral load and relieve many symptoms of HIV disease.

Broad-Spectrum Antibiotics Also Perturb Our Microbiomes

Our bodies contain trillions of bacteria that have evolved to live in humans. More than 38,000 different species live in the human digestive tract, and bacteria occupy at least 20 distinct niches on our skin. The microbes carried by each host are collectively called a microbiome. Humans have evolved to take advantage of the bacteria, and the bacteria gain advantage from us. Box 1-2 describes examples relating to obesity and pain. Some bacteria help humans digest food, whereas others protect from particular pathogens. For example,

Box 1-2: Microbiomes Contribute to Obesity and Pain

Although human digestive tracts contain many different types of bacteria, more than 90% of the total is composed of two general types: the Bacteroidetes and the Fermicutes. These bacteria, along with others, extract energy from foods that would otherwise be indigestible. Obese persons have a higher percentage of Fermicutes in their guts than thin persons, and when obese persons lose weight, the percentage of Bacteroidetes increases. The increased fraction of Bacteriodetes appears to be associated with lower harvest of energy from food.[11] A similar difference is observed with genetically obese mice. The obese mice appear to be better able to extract energy from their food, leaving considerably less energy in their feces. When normal, germ-free mice received gut bacteria from obese mice, they put on substantially more body fat than when given bacteria from normal mice, even though food consumption was the same in the two groups. Could gut bacteria contribute to human obesity? Could a shift in microbiome explain why farmers get better growth from cattle fed low levels of antibiotics as "growth promoters"?

Microbiomes may also contribute to sensing some types of pain, as studies with mice indicate. One form derives from inflammation, a complex immune response involving the balance of small molecules called cytokines. Germ-free mice are deficient in the ability to experience a type of inflammatory pain. Introducing bacteria from normal mice into the guts of germ-free animals brought the sensation of pain to normal levels after 3 weeks.[12] Thus, gut bacteria do more than just help mammals digest food.

acid-producing bacteria in the vagina keep yeast populations in check. The complex ecosystem of the digestive tract protects humans from *Clostridium difficile*, the cause of a serious form of diarrhea and bowel inflammation. An unwelcome consequence of antibiotic treatment is the death of much of our microbiome, which can enable resistant pathogen populations to expand.

Antibiotic Resistance Protects Pathogens

Antibiotic resistance is the capability of a *particular* pathogen population to grow in the presence of a *given* antibiotic when the antibiotic is used according to a *specific* regimen. Such a long, detailed definition is important for several reasons. First, pathogens differ in their susceptibility to antibiotics; thus, pathogen species are considered individually. Second, resistance to one antibiotic may not affect susceptibility to another. This means that the antibiotics must also be considered separately. Third, dose is determined as a compromise between effectiveness and toxicity; dose can be changed to be more or less effective and more or less dangerous. Consequently, the definition of resistance must consider the treatment regimen.

Control of infection caused by a resistant pathogen requires higher doses or a different antibiotic. If neither requirement can be met, we have only our immune system for protection from lingering disease or even death. Indeed, infectious diseases were the leading cause of death in developed countries before the discovery of antibiotics. (They still account for one-third of all deaths worldwide.)

Antibiotic resistance is a natural consequence of evolution. Microbes, as is true for all living organisms, use DNA molecules to store genetic information. (Some viruses use RNA rather than DNA; both acronyms are defined in Appendix A, "Molecules of Life.") Evolution occurs through changes in the information stored in DNA. Those changes are called mutations, and an altered organism is called a mutant. Therefore, an antibiotic-resistant mutant is a cell or virus that has acquired a change in its genetic material that causes loss of susceptibility to a given antibiotic or class of antibiotics.

Antibiotic-resistant pathogens need not arise only from spontaneous mutations—bacteria contain mechanisms for moving large pieces of DNA from one cell to another, even from one species to another. This process, called horizontal gene transfer (see Chapter 6, "Movement of Resistance Genes Among Pathogens"), enables resistance to emerge in our normal bacterial flora and move to pathogens. It is part of the reason that excessive antibiotic use and environmental contamination are so dangerous.

A pathogen is considered to be clinically resistant when an approved antibiotic regimen is unlikely to cure disease. We quantify the level of pathogen susceptibility through a laboratory measure called minimal inhibitory concentration (MIC), which is the drug concentration that blocks growth of a pathogen recovered from a patient. (Pathogen samples taken from patients are called isolates.) A pathogen is deemed resistant if the MIC for the drug exceeds a particular value set by a committee of experts. Clinicians call that MIC value an interpretive breakpoint. Infections caused by pathogen isolates having an MIC below the breakpoint for a particular antibiotic are considered treatable; those with an MIC above the breakpoint are much less likely to respond to therapy. The MIC for a given patient isolate, reported by a clinical microbiology laboratory, helps the physician make decisions about which antibiotic to use. For example, if the isolate is resistant to penicillin but susceptible to fluoroquinolones, the physician may choose to prescribe a member of the latter class.

Resistant microbes can spread from one person to another. Consequently, an antibiotic-resistant infection differs qualitatively from a heart attack or stroke that fails to be cured by medicine: Antibiotic resistance moves beyond the affected patient and gradually renders the drug useless, whereas disseminated resistance does not occur with other drugs. Even resistance to anticancer drugs stays with the patient that developed the resistance because cancer does not spread from one person to another. This distinctive feature of antibiotics means that dosing, suitable effectiveness, and acceptable side effects must be decided by different rules than apply for treatment of noncommunicable diseases. The key concept is that using doses that are just good enough to eliminate symptoms may be fine for diseases such as arthritis, but it is an inadequate strategy for infectious diseases. Nevertheless, that strategy has been the norm ever since antibiotics were discovered.

Antibiotic Resistance Is Widespread

The seriousness of antibiotic resistance depends on perspective. For most diseases, we still have at least one effective drug. If we instantly stopped all resistance from increasing, our healthcare system could continue to perform well. But clinical scientists see resistance increasing and call the situation "dire."[13] For some pathogens, such as MRSA and *Acinetobacter*, physicians are forced to turn to antibiotics abandoned decades ago due to their toxic side effects. Our collective task is to develop attitudes and policies that enable all of us to use antibiotics without causing resistance to increase.

We estimate the extent of the resistance problem by surveillance studies. As pointed out, physicians collect microbial samples from patients and send the samples to clinical laboratories for testing (more than 2 billion per year in the United States[14]). Pathogens are cultured, and their susceptibility to specific antibiotics is determined (described in Chapter 2, "Working with Pathogens"). Surveillance workers then collect the data and calculate the percentage of the cultures that are resistant. (MIC breakpoints are used as the criterion for resistance.) This percentage, called the prevalence of resistance, indicates whether a particular antibiotic treatment is likely to fail due to pre-existing resistance. Surveillance also reveals trends when samples are obtained over several years from a similar patient population. Seeing the prevalence of resistance increase gives health planners advance warning that a change in treatment regimen is required.

Often, the prevalence of resistance is low for many years, and then it increases rapidly (see Figure 1-2). The challenge is to identify resistance problems while prevalence is still low. Then public health measures, such as increasing dose or halting the spread of the pathogen, may stop the increase. Many examples exist in which local outbreaks of resistance have been controlled. However, on a global level no antibiotic has returned to heavy use when resistance became widespread. Instead, the antibiotic is replaced with a more potent derivative.

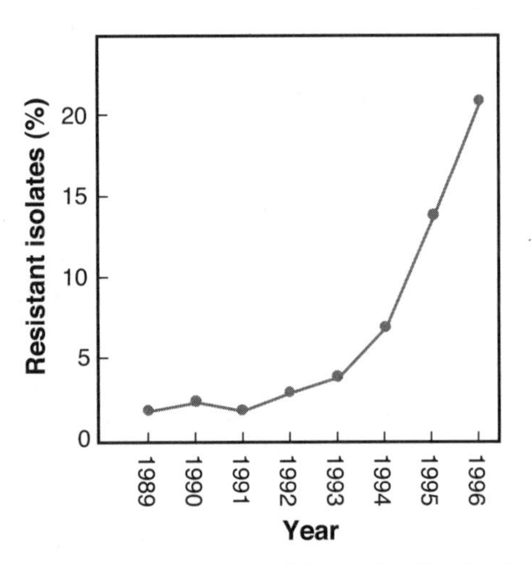

Figure 1-2 Change in prevalence of methicillin resistance in *S. aureus* in Great Britain.

Data replotted from Johnson, A.P. "Antibiotic Resistance Among Clinically Important Gram-Positive Bacteria in the UK." *Journal of Hospital Infection* (1998) 40:17–26.

A partial list of major resistance problems is shown in Box 1-3. This list should be considered as a status report that needs to be continually updated, because pathogens are acquiring resistance to more and more antibiotics. It is also important to point out that resistance is generally a local or regional problem. For example, the prevalence of multidrug resistant (MDR) tuberculosis is particularly high in portions of Eastern Europe and South Africa, but in the United States it is rare.

Box 1-3: Resistance Problems

Several pathogens are close to becoming difficult to treat with antibiotics in some geographic regions. The pathogens and geographic locations listed in Table 1-1 are examples; a comprehensive listing of problem pathogens would require many pages.

Table 1-1 Examples of Pathogens That Have Become Extensively Resistant

Pathogen Species	Disease	Drugs Exhibiting Resistance	Geographical Locations
Acinetobacter baumannii	Pneumonia; wound and urinary infections	All common drugs available; polymyxin is still useful in some localities	Reported worldwide in hospital ICUs[15]; pan-resistant in S. Korea, Thailand[16,17]
Klebsiella pneumoniae	Pneumonia	Carbapenen, fluoroquinolones, amino glycosides, cephalosporins	Hospitals in many countries, New York City, South Florida[18,19]
Mycobacterium tuberculosis	Tuberculosis (XDR-TB)	Rifampicin, isoniazid, fluoroquinolone, second-line injectable (kanamycin, amikacin, capreomycin)	Worldwide, particularly Eastern Europe and South Africa[20,21]
Neisseria gonorrhoeae	Gonorrhea	Penicillins, tetracyclines, fluoroquinolones, macrolides, cephalosporins	Western Pacific, Japan[22,23,24]
Salmonella enterica	Food-borne bacteremia	Ampicillin, chloramphenicol, tetracycline, sulfamethoxazole, trimethoprim, fluoroquinolones	Worldwide[25,26]
Staphylococcus aureus	Many types of infection	β-lactams, fluoroquinolones, gentamycin	Worldwide; examples from European hospitals[27,28]

Antibiotic Resistance Is Divided into Three Types

Antibiotic resistance is categorized into several types that require different solutions. One is called acquired resistance. As a natural part of life, mutant cells arise either spontaneously (about one in a million cells per generation) or from the transfer of resistance genes from other microbes (see Chapter 6). When a mutant is less susceptible to a particular antibiotic than its parent, mutant growth is favored during treatment. Eventually, the mutant becomes the dominant member of the pathogen population. One way to slow this process is to limit antibiotic use or use doses that block mutant growth.

When the "acquired" mutant starts to spread from person to person, it causes transmitted or disseminated resistance. In this second type of resistance, the pathogen is already resistant before treatment starts. Disseminated resistance is often highly visible and may elicit immediate action by the healthcare community. Much of that action is aimed at halting transmission.

A third type of resistance involves pathogen species unaffected by particular antibiotics. They are said to be intrinsically resistant. Little can be done about this type of resistance except to develop vaccines and use good infection control practices that keep the pathogens away from us. Most viruses fall in this category.

The Development of New Antibiotics Is Slowing

For many years, pharmaceutical companies developed new antibiotics to replace old ones whose effectiveness was seriously reduced by resistance. The new drugs were often more potent versions of earlier compounds. Unfortunately, finding completely new antibiotic classes becomes progressively more difficult as we exhaust the available drug targets in pathogens. Early in the Twenty-First Century, pharmaceutical companies placed considerable hope on genomic technology as a way to find new bacterial drug targets and thereby new antibiotics. In this approach, computer-based analyses examine the information in bacterial DNA and gene expression profiles to identify potential targets for new antibiotics. So far, that approach has not panned out. At the same time, pharmaceutical executives realized that more money could be made from quality-of-life drugs and drugs for managing chronic diseases. For example, heart disease requires life-long therapy to lower cholesterol. In contrast, antibiotics are administered for only short times. Antibiotics also have a large development cost, almost $1 billion per drug. As a result, many major pharmaceutical companies shut down their microbiology divisions. Small biotech companies are taking on the effort, but we can no longer depend on new compounds to postpone the antibiotic resistance problem.

Vaccines Block Disease

Vaccines represent an alternative way to combat microbes and viruses. Vaccines are preparations of attenuated pathogen or noninfectious parts of pathogens. When eaten or injected, vaccines create a protective immune response against a particular pathogen. Some vaccines are so effective that they eliminate a disease, as was the case with smallpox. The absence of disease means no resistance problem. Unfortunately, we have been unable to make effective vaccines for many pathogens, most notably HIV, tuberculosis, and malaria. Moreover, pathogen diversity can generate resistance to a vaccine (see Box 1-4).

Box 1-4: Vaccine-Resistant Pathogens

Vaccines typically instruct the human immune system to recognize a pathogen and destroy it. In some circumstances, the pathogen can alter its surface properties to make it less responsive to the immune system. For example, the malaria parasite frequently changes its surface; consequently, the human immune system is always a step behind the parasite. In other cases, the pathogen species exists in many varieties. Shortly after the U.S. anthrax scare of 2001, considerable concern arose because the bacterial strain used in the attacks, the Ames strain, was relatively resistant to the available vaccines.

Vaccines for *Streptococcus pneumoniae* (also known as pneumococcus) illustrate the principle of replacement.[29] This organism, which causes pneumonia, otitis media (middle ear infection), sinusitis, and meningitis, colonizes the nasopharynx of 50% of children and about 2.5% of adults. Two types of vaccine are available, one prepared against polysaccharides of 23 pneumococcal strains and the other against a nontoxic diphtheria protein conjugated to polysaccharide from 7 strains of *S. pneumoniae*. The former reduces the impact of disease, whereas the latter also eliminates colonization by the pathogen. Because more than 90 strains (serotypes) of *S. pneumoniae* have been identified, neither vaccine was expected to provide full coverage. Nevertheless, the 7-strain vaccine reduced invasive pneumococcal disease by more than 70%. The fraction of antibiotic-resistant pneumococci also dropped. However, elimination of vaccine strains as colonizers created an ecological niche for nonvaccine strains. As a result, serotype 19A, which was rare before the vaccine became available, replaced vaccine strains. In some cases, capsular switching occurred between a vaccine strain (serotype 4) and a nonvaccine strain (serotype 19A) due to genetic recombination. The resulting strains have virulence properties of serotype 4 with low sensitivity to the vaccine (serotype 19A).

Another serious example concerns the pertussis vaccine. Before vaccination began in the 1940s, pertussis (whooping cough) was a major cause of infant death. In the 1990s, pertussis began a resurgence in countries where most of the population had been vaccinated. Some of the resurgence was due to waning vaccine-induced immunity among the elderly, who increasingly were stricken with whooping cough. However, in Holland between 1989 and 2004, a new strain of *Bordetella pertussis*, the causative agent, replaced the old one among children, and the number of whooping cough cases increased. The new strain appears to be more virulent and produces more toxin than the old one.[30]

Perspective

Pathogens have attacked humans throughout history. Before the middle of the twentieth century, we relied on our immune systems to survive those attacks. The unlucky and the weak died. Our immune systems were strengthened by improvements in diet, and the frequency of some pathogen attacks was reduced by sanitation and water purification. For other pathogens, vaccines were developed that further decreased the overall burden of infectious disease. Insecticides provided local protection from being bitten by mosquitoes and other disease-carrying vectors. But our fear of pathogens was eliminated only by antibiotics. By taking pills for a few days, we could quickly recover from most bacterial diseases. Resistance is bringing back our fear of the "bugs."

Many of our resistance problems derive from the cumulative effects of several complex factors. One has been our cavalier attitude. For example, in early 2009, American supermarket chains began to advertise free antibiotics to attract customers. The underlying message was that antibiotics cannot be very valuable and worth protecting. Another factor is lack of stewardship. Drug resistance is discussed widely among health officials, but a coherent plan has not emerged. Hospitals are beginning to oversee their own use, but agricultural and community antibiotic use is largely uncontrolled after the drugs are approved by governmental agencies. For years, medical scientists, notably Fernando Baquero, Stuart Levy, Richard Novick, and Alexander Tomasz, wrote and spoke passionately about the dangers posed by resistance. The medical community now uses education as a strategy to limit antibiotic use. As a part of this effort, the Centers for Disease Control (CDC) formulate and distribute plans for restricting the emergence of resistance in particular environments. In one survey, neonatal intensive care units failed to adhere to the guidelines about 25% of the time.[31] Outside hospitals individual patients continue to insist on antibacterial treatments for viral infections, a behavior that stimulates the emergence of resistant bacteria and upsets the balance of microbial ecosystems. In the Latino immigrant community, the prescription process is commonly bypassed.[32,33] Thus, the educational effort needs to be intensified. A third factor is the philosophy behind the choice of dosage. Doses are kept low enough to cause few side effects but high enough to block susceptible cell growth or kill susceptible cells. Conditions that block the growth of susceptible cells but not that of mutants are precisely those used by microbiologists to enrich mutants. Conventional dosing strategies lead *directly* to the emergence of resistance.

Understanding the factors that drive the emergence and dissemination of antibiotic resistance is central to controlling resistance. In the following chapters, we describe how antibiotics are used, how pathogen populations become resistant, and what we as individuals can do about resistance. We begin by considering aspects of pathogen biology relevant to antibiotic treatment.

Chapter 2

Working with Pathogens

Summary: Most pathogens are too small to be seen as individuals with the naked eye; however, on solid surfaces, such as agar, bacteria and yeast grow into visible masses (colonies) containing millions of cells. Many viruses kill the cells they infect, thereby leaving a hole (plaque) in a lawn of host cells growing on a solid surface. Colonies and plaques can be counted to estimate the number of infectious microbes or virus particles present in the cultures. Pathogens can also be detected and identified by nucleic acid hybridization following amplification. Short nucleic acid strands made in the laboratory serve as probes that bind specifically to nucleic acids of particular pathogens. Nucleic acid tests are rapid, specific, and sensitive. Demonstrating that a specific life form is responsible for a particular infection is central to understanding infectious disease. Causality is established by a set of criteria called Koch's postulates. Because new infections continue to emerge, Koch's postulates remain relevant, even though they are more than 120 years old. New disciplines of biology have permitted additional criteria to be considered when investigating causality, which has led to significant modification of the postulates. A key to understanding pathogen biology is the realization that infections often contain huge numbers of pathogen cells; consequently, rare genetic events, such as mutation, occur often enough to be a problem.

Differences among the various pathogens require distinct management strategies, particularly when considering antibiotic therapy. We begin by considering how to detect and count microbes. Then we briefly discuss criteria for establishing causal relationships between putative pathogens and disease. The chapter concludes with a central point for resistance: infections contain large numbers of pathogens that must be considered as heterogeneous populations (populations of susceptible cells containing small subpopulations of resistant mutants).

Pathogens Are a Diverse Group of Life Forms

Pathogens fall into three general types: 1) bacteria, which lack a clearly defined nucleus; 2) fungi, protozoa, and helminths, whose cells have a distinct nucleus and are biochemically similar to human cells; and 3) viruses, which are inert molecules when outside their host cells. Single-celled organisms, such as bacteria, are sometimes thought to lead simple lives: they grow and then divide to form two new cells. Upon closer examination we see that some form spores that permit survival in extreme environments, and many have ways to sense and respond to population density. Fungi such as yeasts are also single-celled organisms, whereas those called molds form filamentous networks and specialized fruiting bodies that produce spores. The spores drift through the air until they land on a suitable nutrient surface. There, they germinate, forming filamentous hyphae. The opportunistic pathogenic yeast *Candida albicans* also generates hyphae upon infection. Protozoa are a third type of single-celled organism. Parasitic protozoa often have complex life cycles in which some forms grow in insect vectors, whereas quite different forms live in our bodies. Helminths are multicellular worms that invade our bodies; pinworm and hookworm are examples. Viruses lack the molecular machinery for independent life, but when they penetrate our cells, they can force the cells to make viral components. Those components assemble to form progeny virus particles that are then released to infect other host cells. A common feature of these diverse life forms is their ability to multiply inside our bodies and form large populations.

Pathogen Numbers Are Measured by Microscopy and by Detecting Growth

To understand and control pathogens, we must have a way to count them—we need to know whether an antibiotic reduces pathogen numbers. When bacteria are placed on a glass microscope slide, stained with a dye, and viewed through a microscope, they appear as tiny spheres or rods, depending on the species. Most bacteria are surrounded by a protective cell wall. The structure of the cell wall and its ability to take up a particular stain separates bacteria into two general types. One group is called Gram-positive and the other Gram-negative in honor of Christian Gram, the inventor of the stain. These two bacterial groups, which have evolved along separate paths, often differ in their response to antibiotics. Fungal cells and protozoa are much larger than bacteria and are easily observed by light microscopy; most viruses are submicroscopic.

Situations exist in which microscopy is used routinely for diagnosis of disease. One example concerns tuberculosis. With this bacterial disease, samples of sputum (mucus and fluids coughed up from lungs) are stained in a way that distinguishes *M. tuberculosis* from other bacteria. This microscopic diagnosis is rapid, low-tech, and inexpensive. Unfortunately, microscopy does not detect all cases of tuberculosis, in part because the sample is small. Thus, other detection techniques, such as culture methods that enable bacteria to reproduce, are also important.

Most microbes are so small that little detail is seen by light microscopy. For detail and to see small pathogens such as viruses, we turn to electron microscopy. Electron beams have a shorter wavelength than visible light, which enables resolution of much smaller objects. Although electron microscopy has been a powerful research tool, the methods are too cumbersome and the instruments are too expensive for routine measurement of pathogen numbers.

Another way to "see" microbes is to allow them to grow and divide on a solid surface, such as agar. The cells pile on top each other, and when millions are present in the same spot, they form a visible colony (see Figure 2-1). Because all cells in a colony derive from a single cell, they represent a clone. We can estimate the number of cells deposited on an agar plate by counting the number of colonies that form.

Liquid cultures of bacteria often contain hundreds of millions of cells per milliliter. Such dense cultures contain too many cells to count as colonies if all were placed on an agar plate and allowed to grow. (The colonies would grow together and form a lawn.) To solve this problem, we dilute the culture before spreading a small drop over the agar surface. By knowing 1) the extent of dilution, 2) the volume of diluted culture applied to the agar, and 3) the number of colonies that form, we can calculate the number of cells present in the original sample. That number is expressed as colony-forming units per milliliter. This measure is used to evaluate antimicrobial action.

Figure 2-1 Bacterial colonies. A small drop of *Escherichia coli* culture was spread (streaked) on a portion of an agar plate. Then a sterile wire loop was drawn across a small region where cells had been placed; streaking the loop across a clean portion of the agar served to dilute the culture so that individual colonies would be seen. The plate was then incubated at 37°C and photographed.

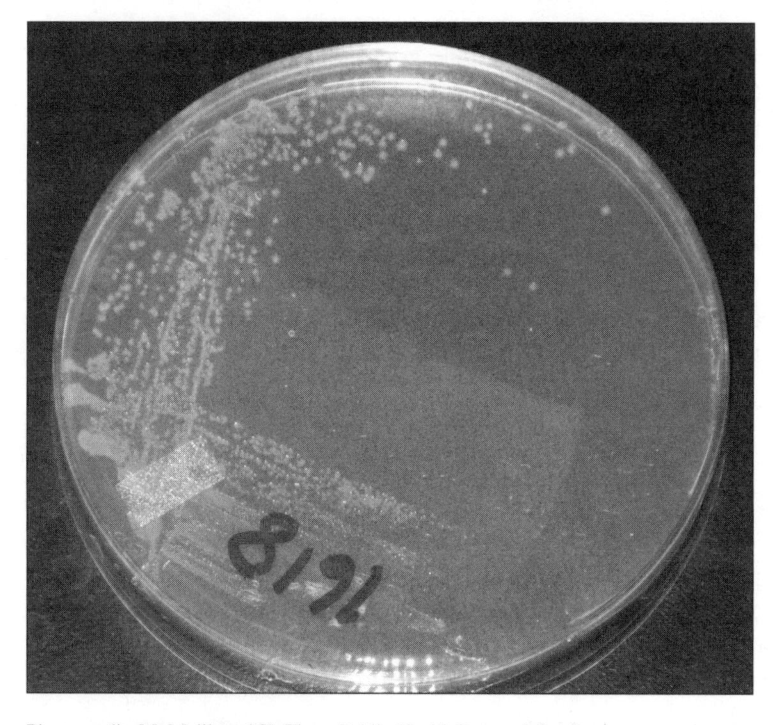

Photo credit: M. Malik and X. Zhao, Public Health Research Institute.

Pathogen culturing methods are frequently used to evaluate antibiotic action, as detailed in Box 2-1. Some antibiotics block pathogen growth, whereas others also kill cells. To measure effects on growth, the antibiotic is present in the culture medium throughout the experiment. To measure lethal action, the microbial culture is exposed to the antibiotic for a specific time, and then the number of surviving cells is determined by growth into colonies on drug-free agar.

Box 2-1: Measuring Static and Lethal Action of Antibiotics

The susceptibility of a pathogen to an antibiotic can be determined in two standard ways by measuring the minimal inhibitory concentration (MIC). One way to assess MIC is by broth dilution. In this method, a series of tubes with liquid growth medium is set up such that each contains a different concentration of antibiotic. Then a small amount of a microbial culture (10^3 to 10^5 cells where $10^3 = 10 \times 10 \times 10 = 1{,}000$) is added to each tube, and the tubes are incubated, usually at human body temperature (37°C). Growth is detected by the medium becoming turbid (cloudy) due to large numbers of cells. The lowest drug concentration that prevents the culture from becoming turbid is taken as MIC. To cover a wide range of concentrations, the drug concentration usually differs by a factor of two between tubes in the series (1, 2, 4, 8, 16, and so on). To allow results from one laboratory to be compared with those of another, standard procedures for determining antibiotic susceptibility have been established by the Clinical and Laboratory Standards Institute (United States) and the European Committee on Antimicrobial Susceptibility Testing. These two organizations are described in more detail in Box 4-2.

MIC can also be determined on solid growth medium by setting up a series of agar plates, each containing a different antibiotic concentration. Microbial cells, usually between 10^3 and 10^5, are placed on the plate. After a suitable incubation period (one day to one month, depending on the pathogen species), colonies form on plates containing low drug concentration but not on those with high drug concentration. The lowest drug concentration that prevents colony formation is taken as the MIC.

MIC measures the capability of a drug to block growth. A different assay measures killing. For this, the pathogen is grown as a pure culture in a test tube or flask, and the antibiotic is added. For kinetic measurements, samples are removed from the culture at specific times, diluted, and applied to agar plates that lack drug. After incubation, the number of colonies is determined and compared with the number obtained from the culture immediately before antibiotic treatment. A few hours of treatment with a lethal antibiotic can reduce colony number by 99.99%.

To measure drug concentration effects, a series of tubes is set up in which antibiotic concentration is varied with incubation time being kept constant. In this assay, the concentration that kills the microbes is called the lethal dose (LD), usually defined in terms of the fraction killed. For example, LD_{90} would be the dose that kills 90% of the microbial population in a specified time. The term minimal bactericidal concentration (MBC) refers to the lowest antibiotic concentration that reduces the number of colony-forming units by 99.9% in an 18-hour incubation (for rapidly growing microbes). It is important to distinguish killing (MBC and LD) from blocking growth (MIC). As pointed out in the text, cells are plated on drug-free media, and the number of survivors is determined for measurements of killing. For inhibition of growth, the drug is present throughout the experiment.

Another variation is called minimal effective concentration, which is used for drugs that alter the morphology of the growing pathogen and render it less pathogenic. An example of this test involves the behavior of the echinocandin antifungals with *Aspergillus fumigatus. A. fumigatus* (see Figure 2-2) is the cause of serious mold infection that can kill cancer and organ transplant patients. The echinocandins do not kill *A. fumigatus.* Instead, they alter its form on agar from a long, filamentous type of growth to a stubby, starfish-like structure, a change that signifies drug action.

Figure 2-2 *Aspergillus fumigatus.* This fungus grows on agar as thin hyphae that form fruiting bodies with spores called conidia.

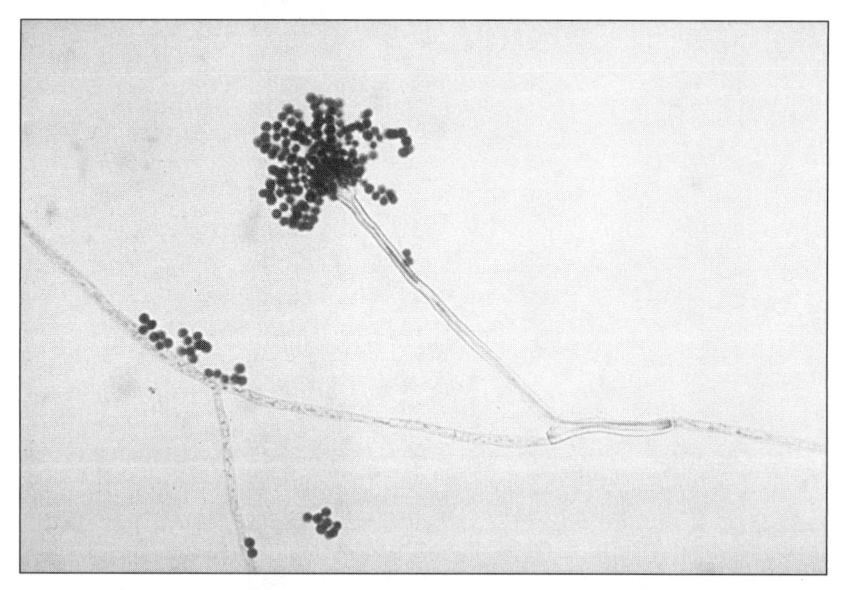

Public Health Image Library #300

The general idea behind counting bacterial colonies also applies to certain viruses. Virus measurement depends on the capability of host cells to grow as a layer on a solid surface. If enough cells are placed on the surface, their growth covers the entire surface. With bacteria, such growth is called a lawn; with human or animal cells, it is called confluent growth (growth of some cell types stops when cells touch each other). If a virus that kills cells is mixed with the host cells *before* a lawn forms or confluent growth occurs, the virus quickly kills the growing cells. Many progeny virus particles are released, and they infect nearby host cells. A zone of cell death spreads over the surface from the point of initial infection. Eventually uninfected cells surrounding the death zone stop growing due to lack of nutrients or contact inhibition. Then the viruses no longer establish a productive infection. The result is a visible hole (plaque) inside a flat mass of host cells (see Figure 2-3). By counting plaques of diluted virus samples, we can estimate the number of infectious virus particles (plaque-forming units) initially present.

Figure 2-3 Detection of bacteriophage. An agar plate is shown on which *Escherichia coli* was spread over the entire plate. At the same time, dilutions of a bacteriophage preparation were placed on the agar as drops. During incubation at 37°C, the bacteria grew into a confluent lawn except where they were killed by the phage. Large clear regions occurred where large numbers of phage were deposited. On the left are small clear zones where only single phage particles were initially present. Each of these multiplied and gave rise to a plaque. (Top row left shows seven plaques.)

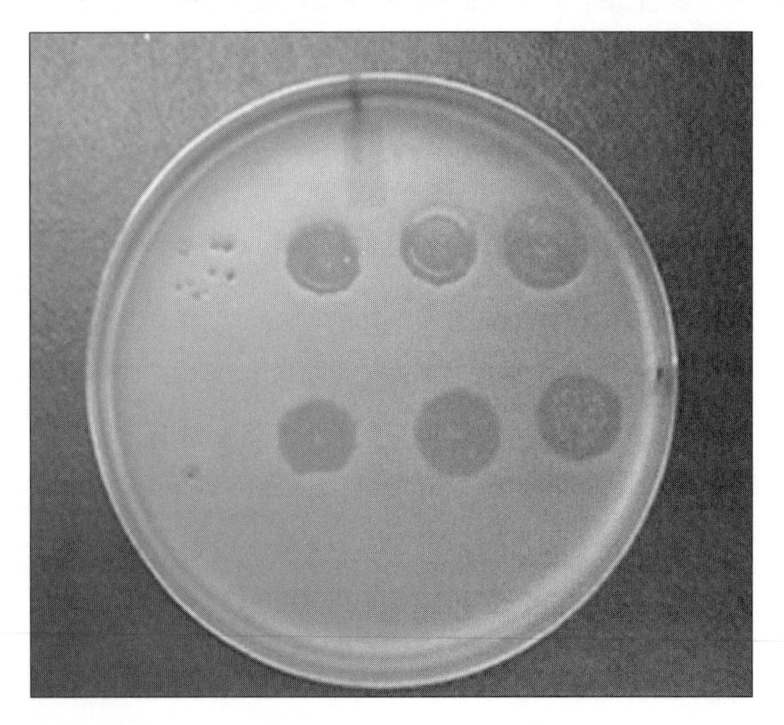

Photo credit: J. Qiao and X. Zhao, Public Health Research Institute.

Not all viruses form plaques. For example, some tumor viruses convert normal human and animal cells into tumor cells that continue to grow and divide after normal cells in the culture stop. In cases where normal cells cover the surface of a Petri dish and stop growing, tumor cells continue to pile on top of each other. The pile of tumor cells is called a focus. Foci can be seen and counted using a low-power microscope. The concentration of tumor viruses is then estimated by knowing the extent of dilution and volumes of virus samples applied to the Petri dish. Some other viruses cause cells to fuse into giant cells called syncytia that can also be seen and counted using a low-power microscope. Thus, the biological properties of pathogens are used to estimate their numbers.

Many pathogens can be handled safely in ordinary microbiology laboratories. To keep from contaminating the cultures, all glassware is sterilized in pressure cookers, as is agar before being placed in Petri dishes. Bacterial cells in colonies are conveniently transferred by touching a colony with the end of a thin wire to pick up some of the cells and then retouching the end of the wire to a clean agar plate or to a liquid growth medium. There, some of the cells fall off the wire and reproduce. To avoid contamination, the wire is sterilized between samples by heating in a flame. Particularly contagious agents are confined to specialized, negative-pressure, biosafety level 3 laboratory rooms where work is performed inside biosafety cabinets. These large, box-like structures have controlled air flow to keep the pathogens inside the cabinet. Workers wear disposable gowns, gloves, and masks. In some cases, full body suits and filtered breathing air is required. As a result of these precautions, laboratory infections rarely occur.

Molecular Probes Can Be Specific and Highly Sensitive

Although tests that require pathogen growth are often easy to perform, they can require considerable time; consequently, physicians frequently prescribe treatment without growing the pathogen and learning the cause of disease. This lack of precision is being corrected by replacement of conventional agar-plate methods with rapid, sensitive nucleic acid tests. For these tests, nucleic acids are extracted from diseased tissue or blood samples, and then they are examined for the presence of a particular pathogen nucleic acid. With DNA, detection begins by forcing apart the two strands of DNA from a patient sample. (Boiling a DNA solution is sufficient to separate the strands, and rapid cooling keeps them from coming back together.) The sample is mixed with a single-stranded DNA probe that is pathogen-specific. Incubation under proper conditions enables the nucleic acid from the laboratory sample, the probe, to bind with single-stranded pathogen DNA obtained from the patient sample. The result is a double-stranded hybrid DNA if the patient sample contains DNA with nucleotide sequences complementary to those in the probe. Formation of duplex DNA containing single-stranded nucleic acids from different sources is called nucleic acid hybridization. Because hybridization occurs only when the nucleotide sequences are complementary, hybridization serves as a specific test for a particular pathogen species. Probes called molecular beacons are available that emit fluorescent light upon hybridization and illumination with visible light (see Figure 8-1).[34]

One problem with nucleic acid hybridization is that the pathogen nucleic acid may represent only a tiny fraction of the total nucleic acid in the patient sample. (Human DNA may be much more abundant than pathogen DNA.) A method called PCR (see Appendix A, "Molecules of Life," Box A-3) enables specific portions of pathogen DNA to be amplified millions of times, making detection much easier and more sensitive. In the case of RNA viruses, such as influenza virus, the viral RNA is first converted into a DNA form before carrying out PCR. This conversion process is called reverse transcription.

We know the nucleotide sequences of many pathogen DNAs; consequently, we can make short complementary DNA probes synthetically. That permits us to bypass work with the living microbe. Synthetic nucleic acid probes are becoming increasingly popular for detecting and identifying pathogens quickly. For example, we can now detect HIV shortly after infection, whereas the old antibody tests required more than one month of infection. Examples exist in which we can even tell whether a bacterial pathogen is antibiotic resistant using nucleic acid probes.[35,36,37]

Koch's Postulates Help Establish That a Pathogen Causes Disease

Determining whether a particular microbe is actually the cause of a given disease is guided by Koch's postulates. In the early days of microbiology, when novel bacteria were regularly recovered from diseased persons and animals, Robert Koch proposed a set of rules to help establish causal relationships. His postulates of 1884 are straightforward:

1. The microbe must be detected in all host organisms suffering from the disease.
2. The microbe must be isolated from a diseased host and grown in pure culture.

3. The cultured microbe should cause disease when introduced into a healthy host.
4. The microbe must be isolated from the inoculated, diseased experimental host and shown to be identical to the original microbe suspected of causing the disease.

Although his postulates are sometimes taken as the definitive test for causality, they were never taken as absolutes, even by Koch. (Postulate 3 used the word "should" rather than "must" because counter-examples were known.) Moreover, viruses, which had not been discovered when the postulates were published, are notoriously difficult to culture (postulate 2), and for some viral diseases we lack an animal model (postulate 3). Nevertheless, there is little doubt about the viral nature of some diseases.

The importance of the postulates is emphasized by a controversy over the viral nature of AIDS. An animal model was not available to establish that the human immunodeficiency virus (HIV) actually causes AIDS, as prescribed by Koch's postulates. In the late 1980s, Peter Duesberg challenged the prevailing idea that HIV causes AIDS (see Box 2-2).[38] If Duesberg were correct, the money being spent to find antiviral agents and vaccines was being wasted. Moreover, the cure for the disease would be changes in behavior and nutrition, not antivirals. Indeed, antivirals used to interrupt the transmission of HIV from mother to a new-born child were said by Duesberg to cause AIDS. Duesberg's ideas were immediately dismissed by the scientific community, sometimes with strong language, and NIH funding for his work was stopped. However, having a well-known scientist cast doubt on the link between HIV and AIDS provided impetus for South Africa to delay treatment of the virus. Delay is thought to have been costly, in part because South Africa was experiencing an epidemic of tuberculosis,[39] a disease that is exacerbated by infection with HIV.

Box 2-2: Koch's Postulates and AIDS

In 1988, Peter Duesberg, a highly respected virologist, stated that "Human immunodeficiency virus (HIV) is not the cause of AIDS because it fails to meet the postulates of Koch and Henle, as well as six cardinal rules of virology."[40] Duesberg emphasized that it had not been possible to detect the virus, provirus, or viral RNA in all cases of AIDS (postulate 1). Moreover, HIV had not been isolated from all AIDS cases (postulate 2). Third, pure HIV did not produce AIDS when injected into animals, and accidental delivery to healthy humans did not always

continues

cause disease (postulate 3). Duesberg's controversial hypothesis was that the many diverse symptoms of AIDS are due to drug use and poor nutrition.[41,42] At the time, an accidental needle stick had been associated with the development of AIDS in a laboratory worker,[43] and AIDS in hemophiliacs was associated with HIV-containing blood transfusions.[44,45,46] Duesberg argued that these infected persons could have developed symptoms from the treatments they received, that HIV was not the cause of symptoms. He attributed an epidemic in Thailand to an increase in testing, not to the presence of the virus.[47]

Challenges to scientific dogmas are part of the scientific process. They move understanding forward by forcing additional experimentation. But they must withstand the sometimes scathing skepticism of other scientists.[48] Because Koch's postulates can be difficult to satisfy, other less direct criteria have emerged to assign causality (see Box 2-3). In the case of HIV and AIDS, good correlations exist between the size of the viral load and AIDS.[49] Data of this type have led to the near universal acceptance of HIV as the cause of AIDS.

Modern Biology Has Refined Koch's Postulates

We continue to face new types of infection as we modify our environment. Diseases once restricted to tree-tops (monkeypox) have come to ground as we chop down forests, Lyme disease spreads with deer in backyards that were formerly woods, and hantavirus jumps to humans when rodents invade homes. Today we are considering abnormal proteins (prions) as agents of transmissible spongiform encephalitis (mad cow disease).[50] Protein-based diseases require new paradigms, because in all other cases the disease-causing agents contain nucleic acids whose replication is easy to understand. Thus, Koch's postulates remain relevant.

Many advances have occurred in molecular and cell biology since Koch's paper. We now have clinical interventions (antibiotics and vaccines) that remove pathogens, and pathogens are known to acquire resistance that overcomes the antibiotics. Both events can correlate with changes in disease, thereby providing evidence that a particular pathogen causes a particular disease. Genomic nucleotide sequence analysis enables molecular ecology and microbial population genetics to contribute to causality arguments that are becoming increasingly complex. Thus, we now have a variety of ways to examine causality (see Box 2-3). At the same time, we are faced with political limitations on use of animals to establish causality (postulates 3 and 4) when a large body of data is already highly supportive. The net result is that a broad approach is used to identify the cause of disease.

Box 2-3: Beyond Koch's Postulates

The importance of Koch's postulates is emphasized by the continuing effort to apply new technologies to the issue of causality. In 1965, Bradford Hill proposed nine criteria to assess the strength of association between a disease and an agent that is thought to cause the disease. (His work revealed the association between lung cancer and cigarette smoking.) None of his criteria, listed here, is absolute, but together they can make a strong case:

1. **Temporal relationship**: The cause must precede the disease.
2. **Strength**: The stronger the association, the more likely it is causal.
3. **Consistency**: Results from different settings should be consistent.
4. **Dose response**: More exposure to the cause increases the risk of disease.
5. **Plausibility**: A theoretical basis should exist for proposing a relationship.
6. **Consideration of alternatives**: The association becomes stronger as alternatives are eliminated.
7. **Experiment**: Alteration of the disease by appropriate changes in conditions.
8. **Specificity**: A particular putative cause is associated with a specific event.
9. **Coherence**: A particular causal relationship should be compatible with existing theory and knowledge.

In 1988, Stanley Falkow modified Koch's postulates to allow genetic analyses to help attribute a disease to a particular microbe.[51] Satisfying Falkow's corollaries, listed next, substantially increased confidence in the identification of disease-causing pathogens:

1. Identify the gene or gene product contributing to virulence of the pathogen.
2. Show that the gene is present in strains of pathogen thought to cause disease.
3. Show that the gene is absent or inactive in avirulent strains.
4. Disrupt the gene and reduce virulence.

continues

5. Introduce the cloned gene into an avirulent strain and restore virulence.
6. Show that the gene is expressed in the infected host organism.
7. Show that a specific immune response to the gene product protects from disease.

Pathogen Studies Focus on Populations

Although infection of individuals can begin with one or a few pathogen cells or virus particles, we are usually concerned with the behavior of large populations. An important property of large microbial populations is that they are not homogeneous, even if they start from a single cell. At any given time, some cells are carrying out different biochemical processes than other cells in the population. Moreover, pathogen populations contain small subpopulations of mutants—between one in a million (10^{-6}) and one in a hundred million (10^{-8}). Mutants are recognized by having properties that differ from those of the bulk population. For example, antibiotic-resistant mutants grow on agar containing antibiotic, whereas wild-type cells do not. Although mutation frequency is a small number, some patients are infected with more than a billion (10^9) bacterial cells or virus particles, as can be the case with pneumonia, abscesses, tuberculosis, and HIV disease. In such cases, resistant mutants are statistically likely to be present before treatment. Resistant subpopulations, which are discussed in Chapter 10, "Restricting Antibiotic Use and Optimizing Dosing," create a fundamental problem for conventional therapy strategies.

Bacterial populations also contain a small number of cells (1 in 100,000) that are not readily killed by antibiotic treatment even though they have normal antibiotic susceptibility. These cells are called persisters. Persisters are not resistant, because these survivors of antibiotic treatment have the same antibiotic susceptibility as the starting population when retested. (Resistant mutants would exhibit low susceptibility.) Bacterial persister cells are thought to be in a semi-dormant state that protects them from antibiotic attack. They constitute an important reservoir of bacteria that can regrow after removal of antibiotic.

Another property of many bacterial populations is quorum sensing, a molecular process by which bacterial cells communicate with each other.

Quorum sensing involves the release of specific small molecules from bacterial cells and the binding of those small molecules to receptors on bacterial surfaces. When a bacterial culture becomes dense, the concentration of the small molecules becomes high enough to trigger a cellular response via binding to the surface receptors. Thus, members of a bacterial population "know" how many others are present.

Finally, many bacteria form dense, structured communities called biofilms, often on surfaces such as teeth, blood vessels, and catheters. Cells in biofilms tend to be less susceptible to antibiotics than the same species growing as a cell suspension. For some bacterial pathogens, quorum sensing appears to be related to biofilm formation. Thus, interrupting quorum sensing might be a way to lower biofilm formation and increase the susceptibility of these bacteria to antibiotics.[52]

Perspective

Working with pathogens often uses simple technology that is easily adapted for detection of resistance. For example, a resistant bacterium will form a colony on drug-containing agar that prevents growth of susceptible cells. Measuring growth can also be simple. When microbes grow and divide in a liquid culture, they eventually become so concentrated that the medium becomes turbid (cloudy). By measuring the turbidity of a culture, we can obtain a rough estimate of the concentration of microbes.

In the late 1990s, the scientific community mounted a massive effort to determine the nucleotide sequence of the human genome. Rapid sequencing methods emerged that were subsequently applied to many pathogens. As a result, we now have complete nucleotide sequences for most of the medically important microbes. The availability of these sequences encouraged the development of many innovative diagnostic strategies based on nucleic acids. Sequence information also enabled the design of nucleic acid-based antibiotics that were expected to be highly specific. By comparison, our current antibiotics are rather crude agents. So far, few successful nucleic acid antibiotics have been developed, largely due to delivery problems. In the next chapter, we describe the major antibiotic classes to provide a context for considering resistance.

Chapter 3

A Survey of Antibiotics

Summary: Chemicals are selected to be antibiotics based on their capability to cure disease with minimal side effects. (The ability to restrict the emergence of resistance has not been a criterion for antibiotic selection.) Chemicals that kill pathogens were first developed by Paul Ehrlich early in the Twentieth Century. Subsequently, Alexander Fleming discovered that secretions from a mold killed bacteria, an observation that led to the discovery of penicillin. Gerhard Domagk found sulfa drugs in the 1930s. Most antibiotics interfere with central life processes, such as synthesis of DNA, RNA, protein, cell membranes, and cell walls. Because human cells also carry out these processes, except for cell wall formation, a challenge is to find agents that act on biochemical properties that differ between pathogen and human. The antibiotics are grouped and named according to similarities in chemical structure. In general, agents that work with one category of pathogen (bacteria, fungi, protozoa, helminths, or viruses) fail to act on others. Some agents attack a broad spectrum of species within a pathogen category, whereas others are restricted to only a few; some kill pathogens, whereas others only block pathogen growth. Antiseptics and disinfectants are general poisons useful for clearing a wide variety of microbes from external surfaces. They are too toxic for internal use.

Each antibiotic has its own features that influence its suitability for a particular pathogen. This chapter introduces the major antibiotic classes. We then use fluoroquinolones to illustrate how antibiotics evolve. We also briefly consider antiseptics and disinfectants, because they help us kill pathogens on surfaces and because their use may contribute to the emergence of antibiotic resistance. If you are unfamiliar with the synthesis of DNA, RNA, and protein, you may find Appendix A, "Molecules of Life," and Appendix B, "Microbial Life," useful when considering processes blocked by antibiotics.

Antibiotics Are Selective Poisons

Antibiotics are relatively small molecules (about 20 to 100 times the size of a water molecule) that interfere with normal life processes of microbes and viruses. Human cells differ enough from pathogens for antibiotics to act selectively. For example, our cells lack walls whereas bacterial cells have them. Consequently, penicillin, which blocks cell wall synthesis, is specific to bacteria. Penicillin has adverse effects, but they arise from other properties. (Some people are allergic to the drug.)

Three general aspects of antibiotics are important when considering effectiveness. First, some antibiotics only *block growth* (static compounds), whereas others also *kill* cells (cidal or lethal compounds). Some drugs are static with one pathogen and lethal with another. For example, rifampicin kills *Mycobacterium tuberculosis* but blocks only the growth of *Escherichia coli* at concentrations usually used. The distinction is important, because static drugs allow the microbes to resume growth when the compound disappears from the body. Fortunately, the human immune system is effective at reducing the number of pathogens in an infection; consequently, static agents, such as tetracycline, can be effective treatments for some diseases. A second important feature is the molecular mechanism of antibiotic action. For example, agents that cause pathogens to break apart (lyse) cause the release of toxic microbial molecules into the patient. Those toxins can lengthen the time needed to recover from disease. A third issue is whether an antibiotic is a broad-spectrum agent or specialized for use with one pathogen species. For most treatment situations, the infecting pathogen is not identified. Treating with a broad-spectum agent allows effective treatment to begin immediately and often without the added expense of diagnostic tests. However, with diseases that require long treatment periods, such as tuberculosis, specialized agents that cause less damage to our normal bacterial flora are preferred. These narrow-spectrum agents are also less likely to select resistant mutants of other pathogens that may co-infect patients. Such benefits are not limited to tuberculosis; consequently, narrow-spectrum agents are likely to become more popular as rapid molecular diagnostic methods become more convenient.

Antibiotics Are Found in a Variety of Ways

Paul Ehrlich, Alexander Fleming, and Gerhard Domagk pioneered the development of antibiotics early in the Twentieth Century (see Box 3-1). Ehrlich made a variety of chemical derivatives that he examined to find ones that worked. Domagk followed in Ehrlich's footsteps with the first agents that were widely used in clinical practice. Their general approach of testing many compounds has evolved into screening procedures that are now applied to hundreds of thousands of molecules. These methods, which are discussed in Chapter 9, "Making New Antibiotics," are built on basic research that identifies potential drug targets.

Fleming found natural antibiotics. To obtain antibiotics from natural sources, samples from those sources are first incubated with a test microbe to determine whether the sample blocks growth. A positive sample is next split into parts with laboratory procedures that separate molecules into different test tubes for analysis. Eventually, the antibiotic molecules are isolated in pure form, and chemical analysis reveals the structure of the active form. Then medicinal chemists increase potency and safety by modifying the structure.

Box 3-1: Origins of Antibiotics: Ehrlich, Fleming, and Domagk

In the late 1800s and early 1900s, Paul Ehrlich, a German physician, focused his attention on dyes that stain animal cells and tissues. Ehrlich wondered why some dyes stained particular tissues but not others. He realized that specific molecular structures produced specific biological effects. He imagined creating a chemical so specific that it would react only with a target pathogen. He hoped to find an agent that would specifically bind to the protozoan that causes sleeping sickness by modifying side chains around an arsenic-containing compound. Ehrlich synthesized a variety of small molecules, and after testing 900 compounds, he had some success. However, the outcome was not spectacular. At about the same time, the bacterium that causes syphilis was discovered. Ehrlich's young colleague, Sahachiro Hata, tested Ehrlich's compound collection against *Treponema pallidum*, the syphilis bacterium. Compound number 606 miraculously cured syphilis, first in animals and then in humans. Compound 606 was the magic bullet Ehrlich sought, albeit for a different disease. The discovery was announced in 1909, and by 1910 compound 606 was available for clinical use. Compound 606, also called salvorsan, and a newer derivative (neosalvarsan) were accepted treatments for syphilis until 1945.

At about the time that Ehrlich was performing his pioneering work, Alexander Fleming was a young physician in London. Salvarsan worked best when injected intravenously, but at the time such injections were difficult. Fleming became proficient with intravenous injections and was one of the few physicians who immediately applied Ehrlich's discovery. Thus, Fleming experienced first hand the value of an antibacterial agent. During World War I, he saw many soldiers die from infection of what seemed to be minor wounds, and for the next decade he thought about ways to stop infections. In 1928, while sorting through a stack of agar plates that had been used to grow *S. aureus*, he noticed that one had a patch of mold (*Penicillium notatum)* growing on top of the bacterial lawn. That in itself was not unusual, but around the mold was a clear zone where bacterial cells had been killed. Fleming quickly realized that the clear zone came from the release of an antibacterial substance from the fungus. He published the discovery and then drifted into other research. A decade later a pair of chemists, Howard Florey

continues

and Ernst Chain, followed Fleming's observation by isolating and synthesizing penicillin. World War II brought a new sense of urgency for anti-infective agents, and by 1944 enough penicillin was available to treat bacterial infections acquired by troops. Fleming, Florey, and Chain shared a 1945 Nobel Prize.

Gerhard Domagk invented sulfa drugs in the 1930s. Like Fleming, Domagk saw many soldiers die from infections in World War I. After the war, he began working for a German chemical company, and in 1932, he discovered that a sulfonamide-containing compound called Prontosil Red cured streptococcal infections in mice. Many derivatives were made, and by 1940 sulfa drugs were standard therapy for pneumococcal pneumonia, childbed fever, and common forms of meningitis. Domagk was awarded a Nobel prize in 1939, but the Nazi government forced him to decline.[53] Sulfa drugs are still used.

Occasionally, an antibiotic is discovered as a byproduct of efforts to find other agents, as was the case with salvorsan (refer to Box 3-1). In the early 1960s, scientists at Sterling Drug Company sought superior versions of quinine, a drug that had long been effective against the protozoan that causes malaria. Among the derivatives they synthesized was a compound that killed Gram-negative bacteria. This substance, called nalidixic acid, was used for many years as a treatment for urinary tract infections. The study of chemical derivatives of nalidixic acid led to the fluoroquinolones, a group of potent antibacterials. With drug discovery, scientists follow many leads that have dead ends. However, some lines of work yield successful products, as briefly described in the following pages and listed in Table 3-1.

Table 3-1 Antibiotic Classes and Resistance Mechanisms

Target Pathogen	Biochemical Target	Antibiotic Class	Examples of Compounds	Resistance Mechanisms[54-65] (a)
Bacteria	Aminoacyl tRNA synthetases	Pseudomonic acids	Mupirocin	Alteration of isoleucy-tRNA synthetase(b)
	Protein synthesis	Oxazolidinones	Linezolid	Alterations in 23S rRNA
	Protein synthesis	Tetracyclines	Tetracycline	Modification of drug structure, efflux
	Protein synthesis	Aminoglycosides	Streptomycin, tobramycin, kanamycin, gentamycin	Alteration of 16S rRNA, r-protein S12; modification of drug structure(b)

	Protein synthesis	Macrolides	Erythromycin, clarithromycin, azithromycin	Alteration of 23S rRNA, r-protein L4; modification of drug/ribosome structure; efflux
	Protein synthesis	Ketolides	Telithromycin	Alteration of 23 rRNA, r-proteins L4, L22
	Protein synthesis	Lincosamides	Clindamycin	Alteration of 23S rRNA; efflux
	Protein synthesis	Streptogramins	Quinupristin-dalfopristin	Alteration of r-protein L22; efflux
	Protein synthesis	Fusidins	Fusidic acid	Mutation of *fusA* (elongation factor G) and plasmid-borne *fusB*, which protects target from drug[(b)]
	DNA topoisomerases	Fluoroquinolones	Ciprofloxacin, levofloxacin, moxifloxacin	Alteration of gyrase and topoiomerase IV; efflux; Qnr[(b)]
	RNA synthesis	Rifamycin	Rifampicin, rifapentine	Alteration of RNA polymerase
	Cell wall synthesis	β-lactams	Penicillins, methicillin, carbapenens, cephalosporins	Alteration of penicillin-binding protein; β-lactamase[(b)]
	Cell wall synthesis	Glycopeptides	Vancomycin, teicoplanin	Alteration of cell wall target, sometimes plasmid borne
	Folate synthesis	Sulfonamides	Sulfanilamide, sulfamethoxazole	Alteration of dihydropteroate synthase
	Folate synthesis	Dihydrofolate reductase inhibitor	Trimethoprim	Alteration of dihydrofolate reductase
	Membranes	Lipopeptide	Daptomycin	Multiple peptide resistance factor
	Mycolic acid synthesis	Isoniazid	Isoniazid	Alteration of catalase, ketoenoylreductase
Fungi	Ergosterol synthesis	Azoles/triazoles	Miconazole, ketoconazole, fluconazole, voriconazole, itraconazole, posaconazole	Amino acid changes in ergosterol and efflux
	Ergosterol synthesis	Allylamines	Tolnaftate, terbinafine	Alteration of squalene epoxidase activity
	Ergosterol	Polyenes	Amphotericin B, nystatin	Increased catalase
	Glucan synthesis	Echinocandins	Micafungin, caspofungin, anidulafungin	Amino acid changes in glucan synthase

continues

Table 3-1 continued

Target Pathogen	Biochemical Target	Antibiotic Class	Examples of Compounds	Resistance Mechanisms[54-65] [(a)]
	Protein synthesis	Flucytosine	Flucytosine	Drug uptake and drug breakdown
	Mitosis	Mitotic inhibitors	Griseofulvin	Undefined
Protozoa	Toxic accumulation of heme in parasite	Quinines	Chloroquine	Mutations in *pfcrt*, a drug transporter gene
	Unknown	Artemisinin	Artemisunate	Pf-ATPase-6
	Glycosome	Sulfonated naphthalene	Suramin	Undefined
	Unknown	Pentamidine	Pentamidine	Undefined
	Protein synthesis	Aminoglycoside	Paromomycin	Drug uptake
Helminths	Binds tubulin	Benzimidazole	Mebendazole	Change in tubulin isoforms
Viruses	Herpes virus replication	Guanosine analogue	Acyclovir	Amino acid changes in viral thymidine kinase
	DNA and RNA replication	Ribavirins	Ribavirin	Undefined
	HIV reverse transcription	Nucleoside RT inhibitor	Azidothymidine, didanosine	Amino acid changes in RT that impair nucleotide incorporation; nucleotide excission[66]
	HIV reverse transcription	Non-nucleoside RT inhibitor	Etravirin	Amino acid changes in RT that impair drug binding[66]
	HIV processing of poly protein	Protease inhibitor	Saquinavir, indinavir	Amino acid changes in protease impair inhibitor binding[66]
	HIV integration	DNA strand-transfer integrase inhibitors	Raltegravir, elvitegravir	Amino acid changes in integrase impair inhibitor binding[66]
	HIV entry into cells	Fusion inhibitors	Enfuvirtide	Amino acid changes in gp41 impair drug binding
	HIV binding to coreceptor	CCR5 antagonists	Maraviroc, vicriviroc	Amino acid changes in gp120[66]
	Influenza uncoating	Adamantanes	Amantadine	Amino acid changes in protein M2
	Influenza neuraminidase	Neuraminidase inhibitors	Oseltamivir	Amino acid changes in neuraminidase

[(a)] Lowered susceptibility to altered drug uptake is common to many agents.

[(b)] Often plasmid- or integron-associated.

Antibacterial Agents Usually Attack Specific Targets

Much of our success with antibiotics has been with antibacterials, partly because bacterial biochemistry differs in many ways from human biochemistry and partly because bacteria are easy to grow and test for antibiotic susceptibility. Next, we briefly describe the major classes according to the processes they block.

Most of our antibacterial agents interfere with protein synthesis, a process that involves the interaction of many macromolecules. Protein synthesis can be quite sensitive to chemical inhibition. For example, the bioterror agent ricin, which is extremely toxic to humans, acts by inactivating ribosomes, the workbenches where proteins are made. Bacterial ribosomes are composed of two subunits (one called small or 30S and the other called large or 50S). The two ribosome subunits, which must come together for protein synthesis to occur, are composed of large RNA molecules (rRNA) and about 50 different proteins. As an early step in the process, more than 20 enzymes (aminoacyl tRNA synthetases) individually connect more than 20 transfer RNAs to 20 different amino acids. Each enzyme recognizes one type of tRNA and joins it to one type of amino acid to form an aminoacyl tRNA. Messenger RNA (mRNA) binds to ribosomes, aminoacyl tRNAs bind to mRNAs attached to ribosomes, amino acids covalently join, and newly made protein separates from ribosomes. Interference with any step can block bacterial growth.

The enzymes that join specific tRNA molecules to their cognate amino acids (aminoacyl tRNA synthetases) are inhibited by a drug called mupirocin. Because mupirocin has side effects when taken internally, it is generally restricted to external use. One of its current applications is clearing *S. aureus* from the noses of healthcare workers and newly admitted hospital patients during efforts to eradicate staphylococci and MRSA from hospitals.

Linezolid is a member of the oxazolidinones, the newest group of antibacterial agents. Linezolid blocks initiation of protein synthesis by binding to the small ribosome subunit, preventing it from joining with the large subunit. The drug is used mainly for treatment of infections caused by Gram-positive pathogens, because linezolid has almost no activity with Gram-negative bacteria. Linezolid received approval from the Food and Drug Administration in 2000, and in 2001 the first resistant mutant of *S. aureus* was reported. Linezolid use is carefully guarded to restrict the emergence of resistance.

The aminoglycosides bind to ribosomal proteins and ribosomal RNA, thereby blocking a variety of steps in protein synthesis. For example, streptomycin, one of our oldest antibiotics, attaches to a protein of the small ribosome subunit and locks the mRNA-ribosome complex in place. That prevents proper movement of mRNA across the ribosome and leads to incorporation of incorrect amino acids into new protein. Other members of this class are kanamycin, gentamycin, and tobramycin.

The tetracyclines first came to market in the 1940s, and over the years they cured many millions of infections. These compounds bind to the small subunit of ribosomes, but unlike streptomycin, tetracycline prevents aminoacyl-tRNA from properly binding. The tetracyclines have been used extensively in agriculture where they are sprayed on trees to cure fire blight and fed to cattle and hogs as growth promoters. One disadvantage of tetracycline is the permanent yellowing of teeth when used with children.

The macrolides are lethal agents that bind to the large ribosome subunit and prevent elongation of the new protein chain. These compounds include the old antibiotic erythromycin and two newer synthetic derivatives, clarithromycin and azithromycin. The macrolides are commonly used for bacterial pneumonia. Azithromycin is noteworthy because it persists for a long time in the body and need not be administered as often as other macrolides. A long antibiotic half-life was thought to be important for patients who have difficulty taking multiple pills each day, and the drug became popular. But long half-life is not necessarily good when resistance is considered, because selective pressure on the bacterial population is maintained for long times.

The large ribosomal subunit is also the target of the lincosamides. One of their members, clindamycin, controls infections caused by the anaerobic bacterium called *Bacteroides*. This normal inhabitant of the human digestive tract can multiply rapidly in tissue damaged by surgery or accident. Because rapid growth of *Bacteroides* causes serious infection, clindamycin occupies an important niche in the antibacterial armamentarium. However, the drug also kills other intestinal bacteria and permits the growth of *Clostridium difficile*. *C. difficile* causes serious intestinal problems, as discussed near the end of Chapter 5, "Emergence of Resistance."

Another essential process in all organisms is DNA replication. The fluoroquinolones are the major inhibitors of this process. The targets of these drugs are two enzymes called DNA topoisomerases. (They change DNA topology by twisting/untwisting DNA and linking/unlinking DNA circles.) As a part of their reaction mechanism, topoisomerases break DNA, pull the broken ends apart, and pass another region of DNA or another DNA molecule through the gap. Then the topoisomerases seal the break. The fluoroquinolones trap the enzymes on

DNA as drug-protein-DNA complexes in which DNA is broken. These complexes act as road blocks for the DNA replication machinery, thereby inhibiting cell division. The compounds appear to kill cells when the ends of the broken DNA are released from the drug-enzyme complexes, thereby fragmenting the bacterial chromosome. Because all bacteria are likely to contain DNA topoisomerases, any bacterium that takes up fluoroquinolone is expected to be susceptible.

RNA synthesis is also an essential process. Rifampicin is the most successful inhibitor of RNA synthesis. In the case of *M. tuberculosis*, rifampicin is highly lethal and serves as a major, first-line antituberculosis agent. Rifampicin is also active with *S. aureus*, and it is being drawn into clinical practice for MRSA. However, resistant mutants arise so often with *S. aureus* that rifampicin is rarely used in the absence of a second antibiotic.

Inhibitors of cell wall synthesis include the penicillins and their more recent derivatives, collectively called β-lactams. Treatment of growing bacteria with penicillin causes the cells to break apart: A turbid (cloudy) culture of susceptible bacteria will become clear. Bacteria that are not growing and making new cell wall material are generally not killed by penicillin. Four β-lactam classes have been developed: penicillins, cephalosporins, carbapenams, and monobactams.

Vancomycin is another cell-wall-synthesis inhibitor. It is an old drug, first approved for use in 1958. Vancomycin was originally isolated from a soil sample containing a species of *Streptomyces* that had been collected by a missionary in the interior jungles of Borneo. Vancomycin initially commanded considerable interest because *S. aureus* only rarely acquires resistance to the drug.[67,68] Then methicillin and other new β-lactams displaced vancomycin, because the latter needs to be delivered intravenously. Eventually, resistance to the new β-lactams became widespread, and *S. aureus* acquired resistance to most other compounds. That led to the resurrection of vancomycin. For some infections caused by MRSA, vancomycin is now the only major agent available.

Daptomycin is another old compound initially obtained from *Streptomyces*. Early clinical trials revealed side effects at multiple, high doses, and the compound was shelved. (In the 1980s many other compounds were still available for *S. aureus*.) Daptomycin was brought back to the market when strains of *S. aureus* and *S. pneumoniae* became resistant to other agents. Daptomycin acts on membranes of Gram-positive cells; it has little effect on Gram-negative bacteria.

The sulfonamides (sulfa drugs) are bacteriostatic inhibitors of the enzyme dihydropteroate synthetase, which accelerates a key step of folate synthesis.

(Folate is necessary for nucleic acid synthesis.) Because mammalian cells do not make folate (humans get folate from the diet), human cells are not affected by inhibitors of folate synthesis. In the late 1930s, a sulfa craze broke out, and thousands of derivatives were made. Some formulations were dangerous and led to the founding of the Food and Drug Administration (see Box 3-2). Sulfa drugs played a central role in preventing wound infections during World War II. (American soldiers were issued sulfa powder and instructed to sprinkle it on open wounds.) Allergic reactions are one of the problems with sulfa drugs; about 3% of the population experiences adverse reactions.

Antibacterial Agents May Have a Generalized Effect

Antibacterials act by binding specific targets and then corrupting specific biochemical processes. However, they also appear to stimulate a bacterial suicide process that amplifies the effect of the antibacterial. Recent work indicates that some lethal action arises through a response of bacteria to lethal stress in which peroxides accumulate and then decay to form toxic

Box 3-2: Sulfa Drugs and the FDA

In the 1930s, journalists and product safety activists, seeking stronger legislative protection, widely publicized a variety of injurious products. Among these were radioactive beverages, cosmetics that caused blindness, and worthless "cures" for diabetes and tuberculosis. For five years, a proposed law was tabled in Congress. At about the same time, sulfa drugs were discovered, and by the late 1930s, hundreds of companies were making tens of thousands of similar compounds in what is called the sulfa craze. In the fall of 1937, a preparation of a sulfa drug, dissolved in the poison ethylene glycol, was distributed as Elixir Sulfanilamide. More than 100 persons died, and the public outcry led to almost immediate passage of the Food, Drug, and Cosmetics Act. The previous law was so weak that the offending material could be seized and pulled off the market only because it failed to conform to the definition of elixir—it contained no alcohol. The new law required safety testing and led to the general idea that antibiotics should have high standards of safety. This law remains the foundation of the Food and Drug Administration regulatory authority.[69,70]

molecules called hydroxyl radicals.[71, 72, 73] Hydroxyl radicals last only a fraction of a second, but during that time they can break nearby macromolecules. Small molecules are available that block formation of hydroxyl radicals. Treatment of *E. coli* with these agents almost completely protects the bacterium from the lethal action of oxolinic acid, an early type of quinolone. Thus, the idea is emerging that lethal signals caused by antibiotics trigger a cascade of reactive oxygen species that are responsible for much of the cell death.

Bacterial cells contain protective enzymes that break down peroxides, and when one of these enzymes is absent, several antibiotic types become more lethal.[73] That raises the possibility for making inhibitors of the protective enzymes that will enhance the lethal activity of many antibiotics. Although such enhancers have not yet been developed, small-molecule additives have proven successful for β-lactams such as penicillin. Some bacteria make enzymes called β-lactamases that destroy β-lactams. When β-lactamase inhibitors are added to β-lactams, as is the case with Augmentin, the β-lactams are much more effective.

Most Antifungal Agents Attack Membranes and Cell Walls

All cells are surrounded by a semi-permeable membrane that contains a variety of proteins, lipids (fats), and sterols. Cholesterol is one of the sterol components of mammalian cell membranes. Fungal cells are biochemically similar to human cells, but their membranes contain ergosterol rather than cholesterol. The enzymes responsible for making ergosterol differ from those involved in making cholesterol; consequently, drugs that interfere with ergosterol formation are specific to fungal cells.

The yeast *Candida albicans* (see Figure 3-1) is a normal inhabitant of the mouth, throat, and vagina; it is kept under control by host defenses. However, *C. albicans* causes life-threatening systemic infections in persons undergoing cancer chemotherapy or otherwise experiencing immune suppression (see Box 3-3). Among the effective antifungals are triazoles, such as fluconazole, that target ergosterol. Triazoles weaken the membrane surrounding fungal cells and prevent fungal growth. Another effective drug is the polyene called amphotericin B. It binds to ergosterol and creates holes in fungal cell walls. Those holes then kill the fungus. Although amphotericin B is more active than triazoles, it has serious side effects that include kidney damage. Amphotericin B is generally not the first choice for fungal treatment.

Figure 3-1 *Candida albicans*. A photomicrograph of a dense culture shows hyphal-like growth characteristic of this organism that also has single, yeast-like cells.

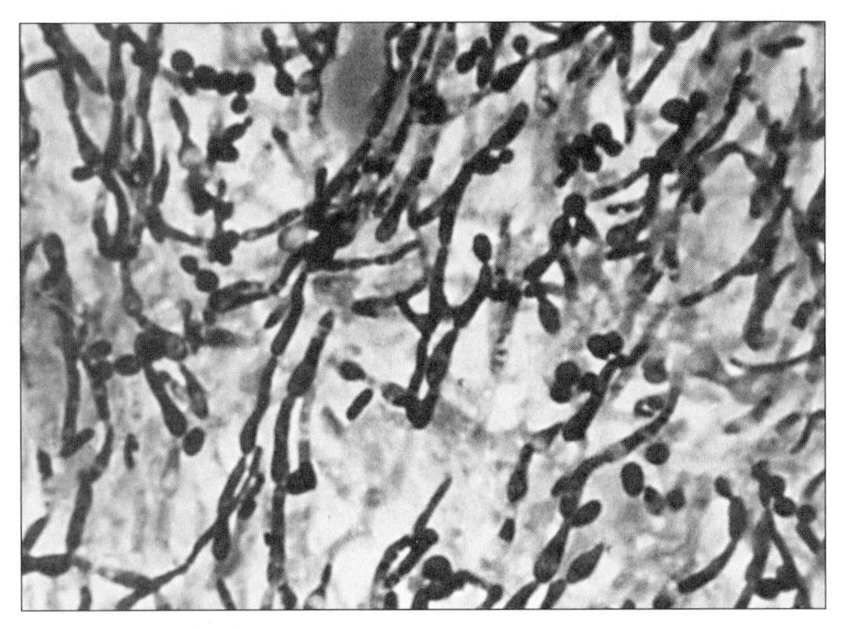

Public Health Image Library #2926

The echinocandins are a new class of antifungal compound that recently came to market in response to growing resistance among triazole antifungals. The echinocandins target an enzyme called glucan synthase, which is responsible for producing the main building block of the fungal cell wall.

Fungi are also attacked by flucytosine. This agent is structurally similar to the nucleotide subunits of DNA and RNA; when flucytosine incorporates into

Box 3-3: Immune Modulators and Fungal Infections

The immune system is powerful, and sometimes it goes astray. For example, rheumatoid arthritis, a painful swelling of joints, involves the excessive action of tumor necrosis factor-α an important protein component of the immune system. Anti-inflammatory agents, such as corticosteroids, relieve the symptoms of arthritis. But they also weaken the defense against fungi. Because the current medicines never actually cure arthritis, they must be taken for life. That makes fungal infection a constant threat.

mRNA, it prevents synthesis of proteins by fungal ribosomes. In addition, enzymes in fungal cells chemically modify flucytosine, causing the drug to inhibit another enzyme (thymidylate synthetase) that is essential for DNA synthesis. Unfortunately, resistance emerges readily. Consequently, flucytosine has limited utility.

Fungal infections range from being a nuisance to being life threatening. A variety of skin infections, such as ringworm, athlete's foot, and jock itch are caused by fungi. Skin infections are often controlled by topical creams containing tolnaftate, an inhibitor of ergosterol synthesis. Tolnaftate and related compounds are generally ineffective against fungal infections that grow under the nails of toes and fingers, because nails serve as tough barriers to drug entry. However, nail infections can be treated with another compound, griseofulvin, which is administered systemically. Many months of treatment are required to permit the nails to grow out fungus-free. Among the side effects of griseofulvin treatment are liver problems; consequently, liver function is usually monitored during treatment.

Antiprotozoan Agents Tend to Be Disease-Specific

Protozoan diseases are considered individually, because the antibiotics used are specific to a particular pathogen. (Many antibacterial and antifungal agents are broad spectrum.) At the top of the list is malaria, which is caused by members of the genus *Plasmodium*. In 2004, the World Health Organization estimated that 300 to 500 million episodes of malaria occur each year and that about 1 million persons die of the disease annually. Malaria is out of control in parts of India and Africa. Quinine, which was discovered to have antimalarial properties by ancient Peruvians (see Box 3-4), served as the main antimalarial agent for many years. Potent derivatives, such as chloroquine and quinacrine, gradually replaced quinine. Chloroquine is thought to act by accumulating in the malaria parasite and causing the buildup of a metabolic product that kills the parasite. Another antimalaria drug, artemisinin (see Box 3-4), is extracted from the wormwood plant, a native of China. Artemisinin derivatives are 10 to 100 times more effective than chloroquine at clearing the malaria parasite from humans. How artemisinin works is not known.

Box 3-4: Ancient Malaria Remedies

Native Peruvians discovered that the bark of the chinchona tree contains medicinal properties for malaria. In the seventeenth century, Jesuits took that information to Rome, where malaria was also a serious problem. The active ingredient of the bark is quinine, a bitter-tasting substance that remained the treatment of choice until the 1940s. For many years, the bark was a significant export from Peru to Europe, and the ground extract was a major factor in the colonization of Africa by Europeans. The Dutch smuggled chinchona seeds out of Peru, and by the mid-1880s, they had established large plantations in Indonesia. Quinine remained the only antimalaria drug until the 1920s when Germans synthesized derivatives. Some of these were captured by Americans in World War II, and subsequent chemical modifications led to chloroquine.

Artemisinin is an ancient Chinese herbal medicine used since the Fourth Century to control fevers, but not specifically malaria. The active compound was isolated in 1971, and it has proved to be an effective antimalarial agent.

Trypanosomes represent another group of medically important protozoa. They cause sleeping sickness in Africa; in parts of South and Central America they cause leishmaniasis and Chagas disease. These diseases are spread by insects and can be fatal, particularly to children. Trypanosomes contain an organelle called a kinetoplast that is absent from human cells. (Kinetoplasts are mitochondria that contain a large mass of circular, interlinked DNA molecules.) The activity of the kinetoplast is blocked by a compound called pentamidine. Trypanosomes are also unusual in containing an organelle called a glycosome where many enzymes involved in the breakdown of sugars are located. Possessing a glycosome makes trypanosomes highly sensitive to a drug called suramin.

The protozoans *Giardia* and *Cryptosporidium* are less life threatening than trypanosomes, but they occasionally cause a form of diarrhea that is difficult to cure. Both protozoans occur in water supplies, with *Giardia* found even in crystal-clear mountain streams. In 1993, *Cryptosporidium* reached the Milwaukee, Wisconsin, water supply and sickened 403,000 persons.[74] Paromycin, which is also an antibacterial agent, interferes with *Giardia* protein synthesis. Another effective compound is metronidazole. This antibiotic is converted into a DNA-breaking agent by enzymes that are active mainly when oxygen concentration is low.

Antihelminth Agents Are Used with a Variety of Worms

Humans carry worms of many types that cause a variety of symptoms ranging from chronic fatigue to death. Helminths are multicellular and have distinct organ systems (for example, muscular, nervous, digestive, and reproductive). Mebendazole is a leading deworming agent. It binds to tubulin, the protein subunit of microtubules, primarily in cells of the worm gut and worm surface. (Microtubules are long structural elements of eukaryotic cells.) Disruption of microtubules leads to reduced uptake of nutrients and starvation of the worms. Another drug, levamisole, is sometimes used with mebendazole. Levamisole paralyzes the worms, and they are expelled alive.

Antiviral Agents Are Often Narrow Spectrum

Understanding antiviral agents often requires knowledge of the particular virus being controlled, because antivirals tend to be specific to one or a few closely related viruses. The many viruses fall into two general categories: those that use DNA as their genetic material (DNA viruses) and those that use RNA (RNA viruses). All viruses, without exception, are obligate intracellular pathogens. Consequently, most antiviral agents must cross the plasma membrane of human cells to access replicating virus. In contrast, many other pathogens grow outside their host cells, making many antibiotics effective without having to enter human cells. This feature means that antivirals may have more safety problems.

Viruses that affect humans bind to proteins on the surface of our cells that serve as viral receptors. Virus particles then enter the cells, some of the viral components separate, and transcription of the viral genome begins. The cellular ribosomes make viral proteins, and replication of the viral genome occurs. (Some viruses replicate in the nucleus of the host cell, whereas other virus types replicate in the cytoplasm.) As viral components accumulate, they assemble into progeny virus. With many viruses, the genes are expressed in a tightly coordinated order to assure that the correct proteins are present at specified times.

A variety of chemical strategies are used to selectively kill viruses. One involves interference with viral replication using nucleoside analogues. To the virus the analogue appears to be a normal component of DNA, but when incorporated into a new DNA strand, the analogue blocks further DNA synthesis. This process, which is called chain termination, also occurs in human cells, but many of these cells are not replicating their DNA, especially in adults.

In general, inhibitors of DNA replication are well tolerated, at least when short-term toxicity is monitored. Another strategy is to prepare short pieces of DNA or RNA that form strong complementary base pairs with viral RNA or DNA. The resulting hybrids block viral replication and in some cases cause cleavage of viral nucleic acids. These agents, which are called antisense oligonucleotides and interference RNA, have good potential as antiviral agents when methods are developed for placing them at the site of infection. A third strategy is to administer proteins that resemble host cell receptors for a given virus. Viruses bind to these decoys rather than the proper cell receptors. Other inhibitors block viral uncoating and the assembly of new virus particles. In principle, five steps of the virus life cycle can be targeted: viral entry, viral uncoating, genome replication, virion assembly, and viral exit.

Ribavirin is an exception to the rule that antivirals are specific to particular viruses—this broad-spectrum agent is active against a variety of RNA viruses. Ribavirin is a nucleoside analog that forms base pairs with cytosine and uracil, creating mutations when viral replication occurs. Accumulation of many mutations incapacitates the virus. Viruses that are not actively replicating are generally not susceptible to ribavirin-type antiviral agents. Among the pathogens normally treated with ribavirin are Lassa fever virus, respiratory syncytial virus, and hepatitis C virus. With the latter, ribavirin is combined with interferon-alpha, a natural antiviral agent made by our bodies.

Foscarnet is another antiviral agent with activity against a variety of viruses. This drug binds to the polymerases of hepatitis B, HIV, and some herpes viruses. Unfortunately, the high doses that must be used cause serious side effects. Next, we describe several virus-specific agents arranged by virus type.

Human Immunodeficiency Virus (HIV)

HIV is an RNA virus that has a complex life cycle with many steps that are vulnerable to chemical interference (see Figure 3.2). The virus enters human CD4+ lymphocytes, cells that are part of the immune defense system, by first binding to a cell surface receptor and co-receptor. When inside the cell, a viral enzyme called reverse transcriptase converts the RNA form of the viral genome into a DNA form. (This reverse flow of genetic information, from RNA to DNA, led to the name retrovirus for HIV and related viruses.) Viral DNA, through the action of another viral protein (integrase), inserts itself into a human chromosome. From its position in a human chromosome, the virus instructs the human cell to make new copies of viral RNA and proteins, which assemble to

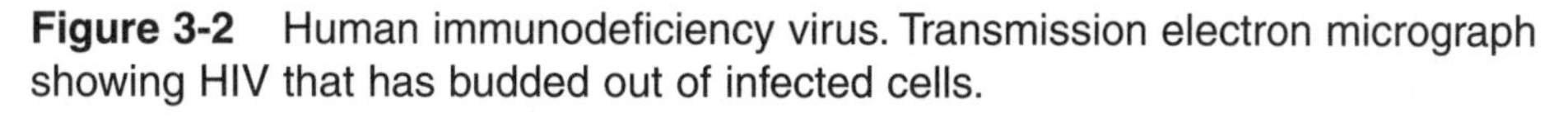

Figure 3-2 Human immunodeficiency virus. Transmission electron micrograph showing HIV that has budded out of infected cells.

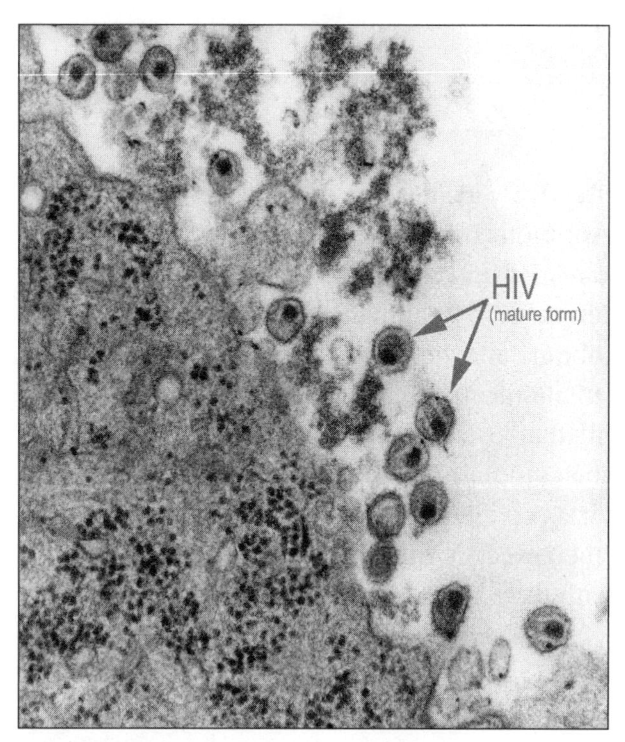

Public Health Image Library # 8254

form immature progeny viruses that move to the edge of the human cell. There, additional viral proteins insert into the cell membrane. Membrane buds containing viral material break off and release virus particles, each surrounded by a coat of host cell membrane decorated with viral protein. One of the interesting features of HIV and related viruses is the production of several proteins as a single, long protein molecule that gets cut into individual proteins by a viral protease. If this cutting fails to occur, the virus cannot reproduce.

Azidothymidine (AZT) was one of the earliest inhibitors of HIV. This agent blocks the action of reverse transcriptase, thereby preventing the virus from converting its RNA genome into the DNA form. AZT is a nucleoside analogue that is incorporated into the growing DNA chain, acting as a chain terminator. Host cells that are actively replicating, such as those lining the digestive tract and some blood cells, are not as sensitive to AZT as the virus. Nevertheless, serious side effects can arise.

Other anti-HIV drugs include several protease inhibitors, non-nucleoside polymerase inhibitors, and integrase inhibitors. Because persons infected with HIV contain large swarms of virus that readily mutate, treatment with a single agent leads to drug-resistant virus. Consequently, persons infected with HIV are treated with a cocktail of several agents.

Influenza Virus

Influenza is a disease for which effective vaccines can be produced, although new ones are needed every year. Antibiotics are useful for persons who fail to mount a strong protective immune response (infants and the elderly). Influenza virus (see Figure 3-3) normally gains access to human cells by binding to the cell surface. Then the plasma membrane attached to the virus retracts (invaginates) into the cell. As the membrane passes into the cytoplasm of the host cell, it forms a hollow ball called a vesicle. Virus particles are inside the vesicles. When a vesicle reaches the cytoplasm, its interior becomes acidic, which enables the outer layer of viral protein to detach from the rest of the virus. The remainder of the virus then escapes from the vesicle and begins to replicate. Amantadine interferes with the removal of viral surface proteins,

Figure 3-3 Influenza virus. Transmission electron micrograph of H1N1 swine flu virions.

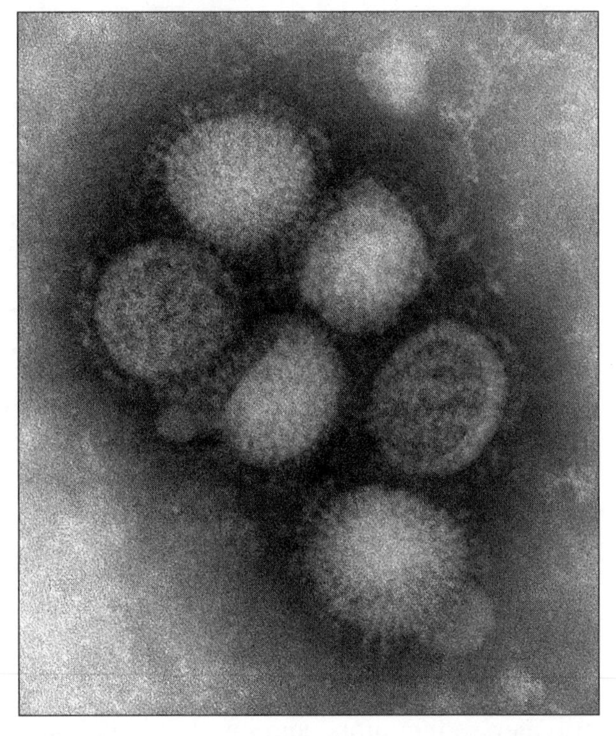

Public Health Image Library #11212; photo credit, C. S. Goldsmith and A. Balish.

thereby trapping the virus inside the vesicles. Oseltamivir (Tamiflu) is a neuraminidase inhibitor that prevents active virus from emerging from infected cells and spreading. (Neuraminidases are enzymes that clip terminal parts off glycoproteins and glycolipids, thereby enabling virus to exit from human cells.)

Herpes Virus

Herpes viruses are large DNA viruses (see Figure 3-4) categorized into eight distinct types that replicate in skin cells where the virus causes lesions. In the case of herpes simplex, the lesions are often called "cold sores" or "fever blisters." These sores heal within a couple of weeks, but they recur, often when a person is under stress. A variety of stressors, such as sunlight, activate the virus. The source of recurring outbreaks appears to be infected nerve cells that harbor the virus in a dormant state and protect it from attack by the human immune system. Acyclovir is an antiherpes agent that acts as a chain terminator, thereby blocking herpes virus DNA replication. In this case, conversion of acyclovir to its active form is carried out efficiently by a viral enzyme but not

Figure 3-4 Herpes virus. Transmission electron micrograph of herpes simplex virus.

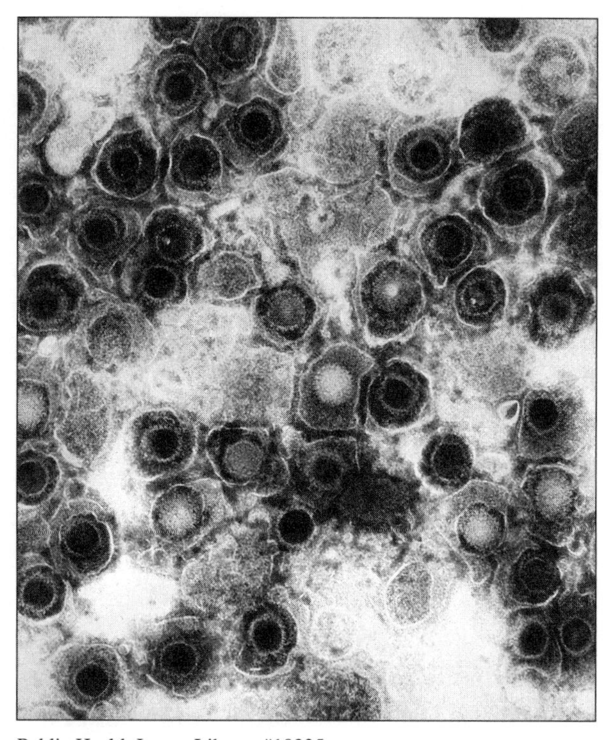

Public Health Image Library #10235

by host enzymes. This selectivity has made acyclovir very useful. Because dormant virus does not replicate extensively, it is unaffected by the drug. Consequently, acyclovir reduces only the symptoms of infection, and then only if taken early in the outbreak cycle. (Herpes drugs do not clear latent infection.)

Antibiotic Classes Evolve

Antibiotic classes evolve when selective forces favor some compounds over others. The selective forces for antibiotics are the intrinsic activity of the compounds, the commercial market (some agents are more successful because their owners have better sales organizations), toxic side effects (side effects can cause a very effective compound to be quickly pulled from the market), and the emergence of resistance. The evolution of the fluoroquinolones serves as an example. Nalidixic acid, the prototype compound, was first reported in the early 1960s. (For structures, see Figure 3-5.) With most bacterial diseases, nalidixic acid is not very effective, and resistant bacteria are recovered from many patients who fail therapy. In the late 1970s, an effort to find more potent derivatives produced norfloxacin. This agent contains a fluorine and several other chemical groups that make norfloxacin effective with Gram-negative bacteria. Within a few years, an even better compound, ciprofloxacin, came along. At about the same time, ofloxacin was synthesized. Ofloxacin is a mixture of right-handed and left-handed forms (mirror images) in which only the left-handed one is active. By purifying the left-handed form (levofloxacin), chemists increased activity by a factor of two. Consequently, levofloxacin replaced ofloxacin in most markets. Ciprofloxacin and levofloxacin proved effective with Gram-negative bacteria. Moreover, both developed excellent safety records, although occasional problems with connective tissue damage kept them from being used with children. (In adults, drug treatment causes an increased incidence of Achilles' tendon rupture, which can be severe.) Both compounds are still among the market leaders.

By the 1980s, the penicillins began to lose their effectiveness with important Gram-positive bacteria, mainly *S. aureus* and *S. pneumoniae*. Ciprofloxacin was tried with MRSA, but resistance emerged quickly: Within a few years, fluoroquinolone-resistant MRSA spread through hospitals worldwide.[75] Ciprofloxacin was also tried in the *S. pneumoniae* market, and again resistance developed.[76] Meanwhile, an effort was mounted to develop new fluoroquinolones that would control Gram-positive pathogens. A compound called trovafloxacin was brought to market in 1998. The compound was highly

Figure 3-5 Chemical structures of quinolone-class molecules.

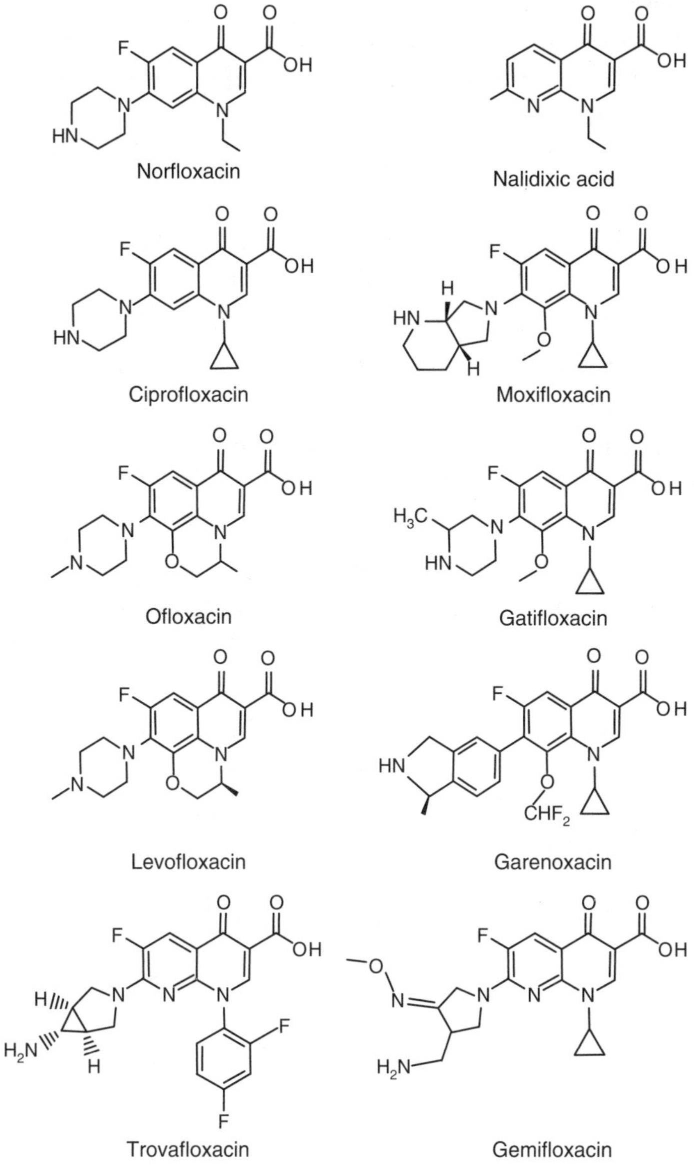

Norfloxacin

Nalidixic acid

Ciprofloxacin

Moxifloxacin

Ofloxacin

Gatifloxacin

Levofloxacin

Garenoxacin

Trovafloxacin

Gemifloxacin

touted, and a year later, it was prescribed at a rate of about 300,000 prescriptions per month. Then 14 patients suffered acute liver failure, and six died. Trovafloxacin was quickly removed from use for most clinical indications, and it became "extinct." As penicillin and erythromycin resistance grew among isolates of *S. pneumoniae*, physicians turned increasingly to levofloxacin, even though it had only modest activity. Levofloxacin had been used safely with millions of patients, and this safety record became a major marketing asset. Soon two other fluoroquinolones, gatifloxacin and moxifloxacin, entered the Gram-positive market. Both were more active with *S. pneumoniae* than levofloxacin, but the levofloxacin sales force continued to increase its market share. In 2006 an analysis of Canadian patients connected gatifloxacin with problems of sugar metabolism.[77] Gatifloxacin was pulled from the pneumonia market, leaving it only as a cure for eye infections. That left the major respiratory market to levofloxacin and moxifloxacin. Two other compounds, gemifloxacin and garenoxacin, came along, but at this writing they have not displaced levofloxacin and moxifloxacin.

A new chapter in fluoroquinolone history is being opened with tuberculosis. Resistance to the two main antituberculosis agents, rifampicin and isoniazid, is growing. Consequently, replacements for these agents need to be found. Because moxifloxacin and gatifloxacin are more active with *M. tuberculosis* than other quinolones, they were chosen for clinical trials to evaluate fluoroquinolone effectiveness. Using a fluoroquinolone for long-term treatment of tuberculosis is problematic, because treatment selects resistant mutants among other bacteria that co-infect or normally live in tuberculosis patients. For example, fluoroquinolone-resistant pneumococci have been attributed to fluoroquinolone use for tuberculosis.[78] The reverse is also true: Treatment of *S. pneumoniae* with fluoroquinolones can lead to fluoroquinolone-resistant *M. tuberculosis*.[79] Thus, adding fluoroquinolones to first-line anti-TB therapies requires consideration of quinolone resistance.[80]

Quinolone-like compounds continue to evolve in research laboratories where efforts are being made to identify derivatives that restrict the emergence of resistance.[81, 82] Without such work, the class is likely to die due to resistance. Resistance has already eliminated fluoroquinolone effectiveness with two major pathogens, MRSA and *Neisseria gonorrhoeae*. With two others, pathogenic *E. coli* and *Pseudomonas aeruginosa*, the prevalence of resistance is approaching the point of serious concern.

Antiseptics and Disinfectants Decontaminate Surfaces

Antiseptics and disinfectants kill microbes, but they are not considered antibiotics because they are too toxic for internal use. They damage pathogens in a variety of ways, from breaking macromolecules to creating holes in cell membranes. Antiseptics are agents applied to the skin. Alcohol and compounds commonly included in hand-washing lotions are considered to be antiseptics.

Disinfectants are agents used on inanimate surfaces. Bleach and formaldehyde fall in this category. Some agents, such as iodine and quaternary ammonium compounds, are in both categories. Antiseptics and disinfectants are especially important in hospitals prior to surgery, because microbes present on skin can enter wounds and establish serious infections.

Perspective

Five factors greatly reduce our fear of pathogens. One is sanitation. By purifying water and protecting food supplies, we reduce deaths to diseases such as cholera and typhoid. Vaccines constitute a second factor. They prime our immune systems to quickly rid our bodies of specific pathogens. The third is good nutrition, because that improves immune status. A fourth factor is insect control. Malaria and yellow fever, once common in the southern United States and Central America, are now uncommon. (Insect-borne diseases are still rampant in Africa.) Antibiotics are the fifth factor. They are especially important when one or more of the first four is absent. For example, antibacterials can be crucial for survival of immunodeficient persons, such as newborns, persons infected with HIV, or patients treated with immunosuppressants. The next chapter discusses how antibiotics are used to cure disease.

Chapter 4

Dosing to Cure

Summary: To respond to disease quickly, initial antibiotic treatments are usually based on symptoms, the physician's experience, and knowledge of diseases currently circulating in the community or hospital. This empiric approach is supported by the availability of antibiotics that control a wide variety of pathogen species. As a backup, pathogen samples (blood, sputum, urine, and so on) are sent to commercial or hospital laboratories for pathogen identification and determination of susceptibility to a variety of antibiotics. (The key susceptibility parameter is the minimal inhibitory concentration, MIC.) Susceptibility information either confirms that the initial treatment was appropriate or indicates a need for change. The antibiotic concentrations that particular doses produce in the bloodstream are determined from animal studies and with volunteers before the antibiotic is placed on the market. The antibiotic concentration usually rises quickly after a dose is administered, and then it drops gradually. The change in drug concentration with time is called pharmacokinetics. To have an effect, the drug concentration at the site of infection should be above the MIC for at least part of the dosing interval. Pharmacokinetic measurements, corrected for pathogen susceptibility (MIC), are used to estimate the effective drug exposure received by the pathogen. Values of exposure can be found that correlate empirically with favorable microbiological results (pathogen eradication) and can be used to define a dose that is likely to give a favorable patient outcome. These predictions are confirmed by clinical studies. To assure that toxic side effects are minimal, clinical studies are also conducted for safety. Governmental agencies then approve particular dosing regimens for general consumption. Duration of treatment is determined by the time necessary to lower the pathogen population to a level that is unlikely to regrow and cause relapse. Preemptive (prophylactic) treatments are often used to prevent infections in surgical patients, because pathogens can easily gain access through wounds. Self-medication, which is often inappropriate (antibacterial agents used for viral infections) or suboptimal (underdosing and inadequate treatment duration), is widespread.

In the 1930s, the public was outraged at the disregard for safety shown by drug manufacturers. As a result, regulatory agencies were established in the United States and Europe. Because drug toxicity is difficult to predict, a guiding principle for many years has been to dose just high enough to cure. In this chapter, we describe how that is done. In later chapters, we show how resistance is forcing a change from the current antibiotic philosophy to one in which doses are chosen to stop resistance without causing harm.

Treatment Strategies Have Been Determined Empirically

We can each recall our physician telling us to "Try this or that, and see how it works." Often, the remedy did work. This strategy, called empiric therapy, is based on trial and error. Guidelines published by the Infectious Disease Society of America (IDSA) and comparable European organizations help in the decision, as do local surveillance studies. Indeed, the physician's guess can be quite accurate if symptoms are distinct or if a particular illness is spreading through the community. Hospital-based clinical microbiology laboratories assist in decision making by providing access to their susceptibility data (antibiograms). When a bacterial infection is suspected, the physician improves the chances of successful treatment by using a broad-spectrum agent, one that is active against a wide variety of bacterial species. Consequently, pharmaceutical companies have focused their efforts on finding and supplying broad-spectrum antibiotics.

For many years, empiric therapy has been an effective approach, largely because most infecting pathogens were susceptible to the therapies and because most patients had strong immune systems that removed microbes. Moreover, empiric therapy can be initiated quickly, which is important with rapidly growing bacteria, and it costs little when no lab test is performed. But the causative agent is not identified, and we learn little about pathogen susceptibility. Consequently, patients sometimes receive ineffective agents. Other negative effects of broad-spectrum agents are also appearing. One arises from blocking the growth of some commensal bacteria, because this allows others to overgrow. We are now beginning to view human bodies as being equivalent to tropical rain forests for microbes, providing a multitude of ecological niches for tens of thousands of different species (see Box 4-1). We cannot always predict effects arising from perturbation of the ecosystem caused by antibiotic treatment.

Box 4-1: Humans as Ecosystems for Bacteria

Many types of bacteria colonize humans. Current estimates indicate that we each carry approximately 10 trillion bacterial cells, a ratio of 10 bacteria for each of our own cells. Biologists estimate that at least 20 different ecological niches exist on human skin alone, each with its own set of commensal bacteria. But the skin bacteria are quite similar: They belong to only 2 of the 70 major bacterial groups. Bacteria in the digestive tract are similar to those found in other mammalian omnivores, constituting some 35,000 to 45,000 bacterial species. Many of the gut inhabitants are killed by treatment with broad-spectrum antibiotics, but they tend to recolonize. That observation suggests that our bodies exert control over the colonizing bacteria.[83]

Susceptibility Testing Guides Antibiotic Choice

Clinical laboratories test samples taken from patients suspected of having a microbial infection. The first task is to identify the pathogen. The traditional approach is to spread the isolate on the surface of an agar plate that is then incubated at body temperature for a day or so. Bacteria and fungi grow on the agar, giving rise to visible colonies whose size, shape, and color help identify the microbe species. Special ingredients in growth media also aid in species identification. More recently, nucleic acid hybridization methods using unique identifiers in rRNA provide rapid, accurate identification. A major problem confronting the clinical microbiologist is determining which colonies on an agar plate are pathogens and which are contaminants picked up during sampling. The problem is particularly difficult with bacteria, such as MRSA, that are sometimes commensal organisms and sometimes pathogens.

The next problem is to determine drug susceptibility for the organism presumed to be the cause of disease. Laboratory technicians test the microbes in the colonies for susceptibility to particular antibiotics, using either the initial isolate, which is done rarely, or cells that grow on the primary agar plates. One popular testing strategy is called the E-test. This test involves spreading a dilute culture of bacterial cells over the entire plate and then placing a strip of paper-like material, impregnated with antibiotic, on the agar before the bacteria start to grow. During incubation the drug leaches out of the paper and blocks bacterial growth. A zone of inhibition results. By varying drug concentration along the paper strip, a point can be seen where drug concentration is too low to cause inhibition. That point correlates with the MIC. A similar strategy, called disc diffusion, involves placing a small paper disc impregnated with antibiotic on the agar. After incubation and bacterial growth, a clear zone is seen surrounding the disc (see Figure 4-1). The diameter of the zone of inhibition is taken as an indicator of susceptibility. MIC can also be measured by placing various concentrations of drug in the agar or by incubating cells in liquid medium containing various concentrations of drug (see Box 2-1).

The MIC or the diameter of the zone of inhibition of the isolate is then interpreted according to sets of guidelines derived from databases compiled by CLSI and EUCAST (see Box 4-2). These agencies of experts collectively designate MIC ranges and inhibition zone diameters as susceptible, intermediate, and resistant. Patients whose isolates fall in the "susceptible" category are likely to experience treatment success. Success is much less likely with isolates that fall in the "resistant" category. The interpretations are reported to attending physicians who use them to determine how to proceed with patient treatment.

Figure 4-1 Disc-diffusion measure of antibiotic susceptibility. A dilute culture of *Escherichia coli* was spread on the surface of an agar plate, and small paper discs impregnated with different antibiotics were placed at various locations on the plate. (The central disc, which is barely visible, contains no antibiotic.) After incubation to enable bacterial growth, a clear zone is seen surrounding each disc due to diffusion of the antibiotic from the disc into the agar. The diameter of the clear zone reflects the susceptibility of the bacterial strain to each antibiotic.

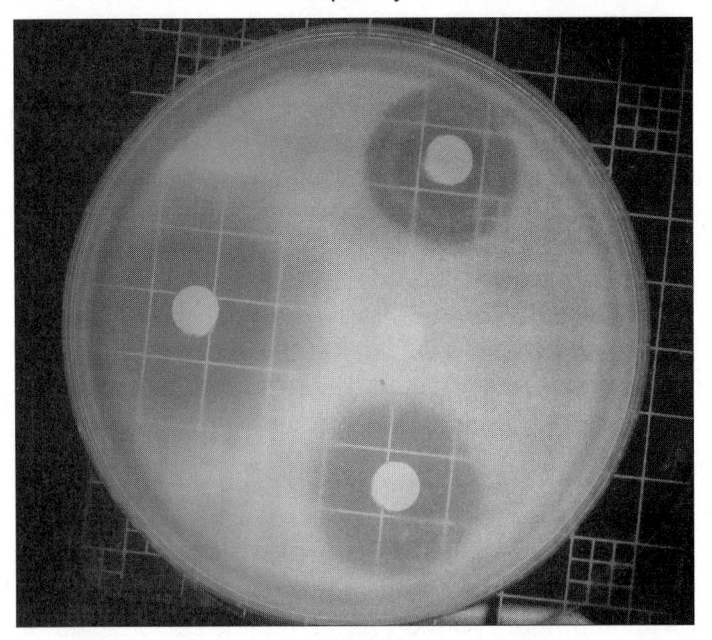

Photo credit: M. Malik and X. Zhao, Public Health Research Institute.

Box 4-2: CLSI and EUCAST

The FDA and its European counterpart, the EMEA (European Medicines Evaluation Agency) set provisional breakpoints for antibiotics during the approval process. Then other groups monitor the susceptibility of pathogen isolates and make recommendations for breakpoint adjustment. In the United States, interpretation is performed by the CLSI (Clinical Laboratory Standards Institute), which was accredited in 1977 as a voluntary consensus standards organization to improve the comparability of medical test results. Similar organizations formed in Europe, and in 2002, European decision making became centralized through EUCAST (European Committee for Antimicrobial Susceptibility Testing).[84] EUCAST receives financial support from government agencies, and its data are freely available on the Internet. CLSI derives part of its financial support from the sale of documents describing standard practices; antimicrobial breakpoints are among the documents.

In many cases, MIC determinations with patient isolates exhibit a bimodal distribution, with wild-type strains and resistant mutants being readily distinguished. However, placement of the "susceptible" boundary can be difficult if isolates of intermediate MIC are present. The goal is to place that boundary high enough so that most susceptible cases receive benefit of a particular drug while treatment failures are minimized. The task is not simple, because humans vary considerably in factors such as drug uptake and excretion. Even the body site of infection is important. Moreover, breakpoints must be reevaluated periodically as new indications for compounds, new treatment regimens, and new methods for testing are introduced. So far, reevaluation has generally caused the breakpoints to be lowered.

EUCAST also maintains a database that lists MIC distributions of patient isolates for various compounds with a variety of microbial pathogens. From these distributions, EUCAST derives epidemiological cut-off values that provide a benchmark against which hospitals or other institutions can compare their isolates for purposes of outbreak control. These values are useful for early detection of resistance (defined as MIC above that seen with the majority—that is, more than 99%—of wild-type isolates). Isolates with such values may or may not be treatable. The database also lists organisms intrinsically resistant to particular antimicrobials.

CLSI and EUCAST differ in several ways. CLSI is composed of representatives from the medical community, from the pharmaceutical industry, and from regulatory agencies; its decisions are made by vote. EUCAST is composed of representatives of national breakpoint committees and the medical and academic communities; its decisions are made by consensus. EUCAST tends to be more conservative with its breakpoints, which are often lower than those determined by CLSI.

Other databases are beginning to emerge as the problem of resistance grows. For example, the Australian Society for Antimicrobials has set up a database called ANZCOSS that specializes in *S. aureus* infections.

Commercial microbial analysis systems are used by most clinical laboratories in industrialized countries. The one called VITEK serves as an example. In this system, the patient sample is loaded into small wells in disposable plastic kits (Test Cards) about the size of a playing card. The 30 wells of each Test Card contain biochemical reagents or antibiotics used to analyze the patient sample. After the wells are sealed, the Test Cards are inserted into an instrument that incubates the cards and optically scans them to determine pathogen species and antibiotic susceptibility profile.

The bacterial culture used for testing, often called an isolate because it is isolated from a patient, is frequently saved in collections as living, frozen samples. The results of testing from many laboratories are collected to provide an overview of the resistance situation. Data in Table 4-1 illustrate how susceptibility information can be used to treat MRSA empirically, that is, without sending a culture to a laboratory. In this example, one would not want to treat with erythromycin if the patient were from Chicago, the city where the survey was taken. Choice of antibiotic is also influenced by whether the person is an outpatient (likely to be infected with community-associated MRSA) or an inpatient (likely to be infected with hospital-associated MRSA). With the latter, multidrug resistance shows a much higher prevalence.

Workers in clinical laboratories usually choose a few colonies to test for susceptibility, which is often sufficient because infections tend to be clonal. Subpopulations of resistant mutants may be present, but they generally go undetected by methods in current use. Consequently, a gradual increase in resistant subpopulation size would not be noticed. That flaw in the method permits resistance to gradually build unnoticed.

Table 4-1 Prevalence of Resistance with MRSA in Chicago*

Antibiotic	Prevalence of Resistance (Percent)			
	Community-Associated MRSA		Hospital-Associated MRSA	
	Children	Adults	Children	Adults
erythromycin	87	93	96	94
clindamycin	7	52	75	74
ciprofloxacin	11	62	62	87
gentamycin	1	11	37	14
tetracycline	6	20	8	13

* Study included all inpatient and outpatients at University of Chicago Hospitals, November 2003–November 2004.[85]

Some bacteria, such as *M. tuberculosis*, grow slowly. (Almost a month is required to form large colonies on agar plates.) Consequently, other diagnostic strategies are sometimes used, such as microscopy, chest X-rays, growth of liquid cultures in special medium, and nucleic acid tests. Setting up a tuberculosis testing infrastructure in developing countries is a challenge. In Box 4-3, we recount experiences in Peru that illustrate efforts involved in upgrading a tuberculosis control program. Whether comparable results can be obtained in other developing countries remains to be seen.

Box 4-3: Developing an Antituberculosis Program in Peru

Peru has one of the highest frequencies of tuberculosis in Latin America (108 per 100,000 persons in 2005; for comparison, the number was 4.5 for the United States). Millions of Peruvians have moved from their highland homes to shanty towns surrounding Lima and other cities. Within these crowded conditions, transmission of tuberculosis is high. In the early 1990s, Peru mounted an intensive treatment program. Significant progress was made, but the program was not designed to stop multidrug-resistant (MDR) tuberculosis. Consequently, the incidence of this form of disease increased.

In 1996, clinicians from several international and Peruvian organizations began to establish an infrastructure for treating all Peruvian tuberculosis patients, including those with MDR disease. Among the problems were difficulties in diagnosis and drug susceptibility testing, especially for MDR tuberculosis. At the beginning of the program obtaining test results required 5 months. In the meantime, physicians would proceed to prescribe drugs empirically. When their guesses were wrong, additional resistance mutations could be acquired by the pathogen. These mutations made the test results inaccurate by the time they came back to the primary care physician. Part of the upgrade effort required methods for quickly and safely transporting bacterial isolates to testing laboratories. Another part was finding ways to speed reporting after the data became available. To solve some of these problems, testing was decentralized. That meant setting up local laboratories. Such labs needed biosafety cabinets for safe handling of the bacterium. However, Peru had no one to certify the proper functioning of the cabinets. Consequently, additional training programs had to be initiated. Moreover, gaining governmental approvals for constructing safe laboratories had significant waiting periods. Nevertheless, the medical will was maintained, and from 1996 to 2000 Peru expanded its capacity to determine drug susceptibility from 50,000 cultures per year to 120,000. The new level has been maintained.[86]

The infrastructure effort is beginning to pay off. From 1997 to 2007, almost 2,000 cases of MDR-tuberculosis were detected and tested for susceptibility to second-line agents. Of these, 119 were extensively resistant (XDR)-tuberculosis. Cure was achieved with 46% of the XDR cases. (Median treatment time was 43 months.) In cases where drug susceptibility testing was performed before the start of treatment, the cure rate was an amazing 71%. (The test results increased the chance

continues

that appropriate antibiotics were selected.)[87] These results, which are better than in most industrialized countries, arose from an intense effort spearheaded by Partners in Health. The situation in Peru serves as a paradigm for disease management and is now being watched to determine whether Peru can sustain such a good record.

Testing for Viruses Bypasses Pathogen Growth

Although many viruses can be grown in the laboratory, the methods are too cumbersome for large-scale commercial testing. Consequently, detection of viruses often uses immunological or biochemical tests. For example, an early test for HIV infection determined whether a patient's body had made antibodies against the virus. A more direct test measures the presence of viral nucleic acid using one of many nucleic acid hybridization methods. With the development of antiviral agents has come the emergence of drug-resistant viruses. If the mutation responsible for resistance is known, viral nucleic acids from the patient can be examined for both the presence of the virus and the nucleotide sequence alteration associated with resistance.

PK/PD Indices Help Determine Antibiotic Dosage

As a part of the licensing process, the U.S. Food and Drug Administration and the European Medicines Agency approve particular doses for each antibiotic. These doses are usually proposed by the manufacturer with support from laboratory tests, animal studies, and clinical trials. Two general properties are paramount: effectiveness and safety. Because not all toxicity problems can be identified using animal tests and small clinical studies, companies have tended to keep doses low, hoping to minimize toxic side effects. Part of the challenge has been to find doses likely to be effective before expensive clinical trials are carried out.

Efforts to identify effective antimicrobial doses use analyses involving pharmacokinetics and pharmacodynamics, commonly abbreviated as PK/PD. Pharmacokinetics describes drug concentration changes that occur in our bodies during therapy. Drug concentration generally rises quickly after we take a dose, and then it gradually drops (see Figure 4-2). When we take the next dose, the concentration again rises and gradually drops. This behavior is described quantitatively in two ways. One is to measure the maximum drug concentration

Figure 4-2 Pharmocokinetics. A single dose of levofloxacin (500 mg) was administered.

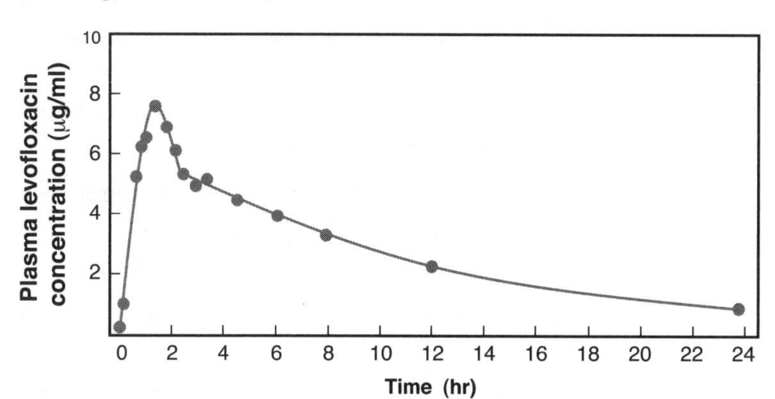

Data replotted from Galan-Herrera, J., Poo, J.L., Rosales-Sanchez, O., et al. "Bioavailability of Two Oral Formulations of a Single Dose of Levofloxacin 500 Mg: an open-Label, Randomized, Two-Period Crossover Comparison in Healthy Mexican Volunteers." *Clin Ther* 2009; 31:1796–1803.

achieved, commonly abbreviated as C_{max}. Another is to draw a graph of concentration versus time and then measure the area under the concentration-time curve (AUC) over a specified time, such as 24 hours. This area is abbreviated as AUC_{24}. Both C_{max} and AUC_{24} change when the dose is changed and both serve as a way to relate dose to drug concentration in the patient.

The effectiveness of an antibiotic dose differs from one pathogen isolate to another because each has a different susceptibility (MIC). Thus, the pharmacokinetic values (C_{max} and AUC_{24}) need to be adjusted for each isolate to estimate the drug exposure received by a particular isolate. This is done by dividing the pharmacokinetic factors by MIC, which produces the expressions C_{max}/MIC and AUC_{24}/MIC. C_{max}/MIC and AUC_{24}/MIC are called PK/PD indices; they reflect antibiotic exposure experienced by the pathogen.

Experimentally, an effective dose can be determined by gradually increasing the antibiotic dose given to infected animals until a dose is found that clears infection in most cases. Pharmacokinetics can be measured with the animals for each dose, which produces values of C_{max} and AUC_{24} that correspond to clearance of infection. When these values are corrected for MIC of the test pathogen, we can obtain a PK/PD index that corresponds to clearance of infection. That value is called a PK/PD target. If this target can be reached when humans are treated, it is likely that infection will be controlled and cleared, assuming that host defense systems are similar in humans and in the test animal.

Although PK/PD indices can be measured for individual patients as a way to obtain the optimal dosing regimen, that is generally not done due to expense and the time needed to obtain the necessary data. Instead, efforts are made to use PK/PD targets to identify doses likely to be effective for human *populations*. With human populations, two variables must be considered. First, patients differ considerably in drug pharmacokinetics for a given dose. (A given dose can achieve different concentrations in different patients.) Second, pathogen isolates from different patients differ in susceptibility: Extensive use of antibiotics has favored the emergence of a variety of mutants, some having reduced drug uptake, some having increased drug efflux, and some having a variety of both. These two variables are handled mathematically to determine how effective a particular dose is likely to be. Operationally, the question is what fraction of the population is likely to achieve PK/PD values, for example AUC_{24}/MIC, that are above the PK/PD target associated with clearance of infection (see Box 4-4).

Box 4-4: PK/PD Indices for Estimating Effective Dose

Patients differ with respect to the amount of drug present in the body for a given dose and in the susceptibility of the infecting agent. Consequently, finding the minimal dose likely to cure most patients is not straightforward. One strategy is to determine a pharmacodynamic (exposure) target (AUC_{24}/MIC) for obtaining cure using a laboratory animal infection model. Then pharmacokinetic measurements (AUC_{24}) are made with large numbers of persons using a given dose, and susceptibility measurements (MIC) are made with a large number of pathogen isolates. Then these two numbers are combined mathematically using a procedure called a Monte Carlo simulation. The output is an estimate of the fraction of patients expected to achieve the target PK/PD index (exposure) for the particular dose. (If the target represents favorable outcome, the Monte Carlo simulation indicates the percentage of treated patients likely to experience the favorable outcome.) That percentage of patients can be increased or decreased by changing the dose.[88] The fraction of patients likely to reach the pharmacodynamic target, for example, likely to be successfully treated, can be balanced with toxic side effects associated with particular doses. Corrections may be needed for immune system differences between the experimental animal models and humans.

PK/PD analyses have led to antibiotics being grouped into two types, those for which higher concentrations kill more pathogen cells (cell death correlates with C_{max} and AUC_{24}) and those for which longer times above MIC give more killing. Members of the first antibiotic type are called "concentration-dependent killers." They include such agents as fluoroquinolones and aminoglycosides. Members of the latter group are called "time-dependent killers." They include β-lactams, such as penicillin.[89]

PK/PD indices are affected by a variety of complex factors, one of which is protein binding. When drugs are inside the human body, they bind to tissues and serum proteins. This binding may sequester the drug and reduce its effective concentration. Consequently, attempts are made to estimate "protein binding" so that a correction factor can be added to PK/PD indices for comparison of doses and compounds. With some compounds as much as 90% is considered to be protein bound, leaving only 10% of the total to control infection. Interpretation of the phenomenon is complex, because it may differ among locations in the body. Moreover, as drug concentration drops during the dosing interval, some bound drug may become unbound. In such cases, protein binding would have created a reservoir of drug. The importance of protein binding, which is controversial, can be estimated by measuring the efficacy of various antibiotics in animals and then determining whether those data fit better with *in vitro* data if corrected for protein binding.[90]

Young Children Are Not Little Adults

For decades, physicians and pharmaceutical companies have treated young patients as small adults when considering antibiotic dosing: Drug doses for children were simply adjusted for body weight. However, metabolism of drugs is strongly age-dependent. In infants, a highly elevated metabolism can result in rapid processing and elimination of drugs. Moreover, drug absorption in infants may be less effective than in adults. Absorption from the gastrointestinal tract is affected by gastric acid secretion, bile salt formation, gastric emptying time, intestinal motility, and microbial flora. All are reduced in infants. However, reduced gastric acid increases bioavailability of some drugs, such as penicillin.

Further complexity arises from consideration of the cytochrome P-450 (CYP450) enzyme system in the small bowel and liver. This system is responsible for much of the breakdown of drugs. CYP450 works in two general ways: 1) oxidation, reduction, and hydrolysis (phase I metabolism) and 2) hydroxylation and conjugation (phase II metabolism). Phase I activity is reduced in neonates, increases progressively during the first 6 months of life, and can exceed adult rates within the first few years. It slows during adolescence, and usually it reaches adult rates by late puberty. Clearly, the effective concentration of drugs in young children should not be based on adult data.

For many years, the U.S. Food and Drug Administration (FDA) considered it unethical to conduct clinical trials on children. Consequently, the small-adult logic was applied. This type of dosing strategy resulted in severe under-dosing of pediatric populations. The consequence has been clinical failures and emergence of drug-resistant pathogens. Fortunately, recognition of the problem is emerging among pharmaceutical companies, regulators, and critical care physicians. Drug dosing in children less than 12 years should always account for age, and adjustment of doses should be based on measurements of plasma drug concentration.[91,92]

Toxic Side Effects Are Determined Empirically

Toxic side effects are often difficult to predict solely from chemical structures, although some chemical groups, such as chlorines and fluorines, are often associated with toxicity. Instead, potential side effects are commonly determined experimentally by trial and error. Initial studies are performed with cultured human cells, and then animals are treated and examined. Finally, human volunteers are tested. Human trials usually involve hundreds to a few thousand persons; consequently, rare safety issues are not seen. The uncertainty of toxic side effects is an important factor in keeping doses low: A few patient deaths, even after a million successful treatments, can be enough to cause a compound to be withdrawn from the market. Consequently, consideration of both efficacy and safety leads to dose recommendations that are compromises—high enough to cure a large fraction of the patient population and low enough to do little immediate harm to most patients. As we point out in Chapter 5, "Emergence of Resistance," this compromise dosing philosophy contributes to the emergence of antibiotic resistance.

An example with HIV and tuberculosis illustrates how some toxic effects come as a surprise. Antiretroviral drugs work well even after HIV has become firmly established; from the HIV perspective, treating immediately after infection with antiHIV antibiotics is unnecessary.[93] Not only are the drugs expensive, but they also have side effects. However, when the patient is also infected with *M. tuberculosis*, a delay in HIV treatment can be catastrophic because HIV frees the bacterium from immune control. If treatment for HIV is delayed, tuberculosis can progress rapidly. But as the antiretroviral drugs enable the immune system to rebuild, the presence of *M. tuberculosis* can lead to a severe inflammatory response called immune reconstitution inflammatory syndrome, which usually occurs in the first month of antiviral treatment. Thus, patients with both tuberculosis and HIV disease must be carefully monitored.

Duration of Treatment Is Determined Empirically

Soon after a patient begins taking an antibiotic, the pathogen population decreases rapidly. Often, the patient quickly feels better. However, a sizable pathogen population may remain if treatment is stopped too early. The remaining pathogens can regrow and cause relapse. In principle, clinical trials can determine the length of time an antibiotic should be taken to minimize the chance of relapse. These expensive experiments have had low priority because the standard 10-day treatment is thought to be adequate.

The damage some antibiotics cause pathogens can last beyond the time that the antibiotic concentration is above the MIC. This phenomenon is called the post-antibiotic effect. For example, fluoroquinolone action appears to break bacterial chromosomes. If repair of those breaks is slow, the bacterial cells will fail to regrow immediately after the drug drops below inhibitory concentrations. Such effects may extend the effective time interval between antibiotic treatments and permit treatments to be given less often. (Antibiotics are generally administered in a series of doses; concentrations rise sharply after each dose and then fall until the next dose is administered.)

Prophylaxis Preempts Disease

Sometimes, antibiotics are administered to eliminate an infection before it causes symptoms or even before an infection develops. For example, surgical patients often receive a low dose or a short course of antibiotic to prevent infection from pathogens that commonly contaminate surgical wounds. Reduced treatment is chosen to minimize toxicity, both as a direct effect of the antibiotic on the host and as an indirect effect of the antibiotic on beneficial microbes. (A full antibiotic course is not deemed necessary because the number of pathogens likely to be present is not nearly as high as in the case of a confirmed infection.)

Prophylaxis is also used with tuberculosis but only after signs of infection. When infected with the tubercle bacillus, only 10% of otherwise healthy persons develop active disease. However, we know that infection occurred in many people without disease, because they exhibit a positive reaction to a TB skin test. (The presence of the bacterium causes an immune response.) In these persons, the bacterium appears to have been forced into a dormant state. Because a subsequent loss of immune function is likely to permit the dormant bacteria to grow and cause active disease, the medical community deems it prudent to treat persons exhibiting a positive skin test, even in the absence of disease. The number of infecting *M. tuberculosis* cells is expected to be low, which has led to the idea that single-drug therapy would be adequate. (Tuberculosis therapy usually involves a four-drug cocktail.) Prophylatic isoniazid treatment is commonly administered for 9 months.

Surveillance studies show that the number of *M. tuberculosis* isolates resistant only to isoniazid is 20 times higher than the number resistant only to rifampicin,[94] one of the other primary antituberculosis agents. (Recall that multiple agents are used with tuberculosis.) Is this an example of prophylaxis causing resistance? Other explanations for greater isoniazid mono-resistance exist. For example, many patients experience side effects from rifampicin and secretly avoid taking the rifampicin pills. That leaves isoniazid more vulnerable to the emergence of bacterial resistance. Isoniazid also has a higher mutation frequency *in vitro*, but the relevance of that has been questioned.[95] At present, we do not know whether prophylaxis contributes to the emergence of resistance.

Management Programs Control Hospital Antibiotic Policy

Antibiotics are used in three general settings: hospitals, community, and agriculture. Only hospitals are starting to take responsibility for stewardship. (The Food and Drug Administration is responsible for efficacy and safety, not drug longevity.) Antibiotic management teams now guide prescribing policies of most U.S. hospitals (see Box 7-1). Among their activities is approving only certain antibiotics for hospital use. These antibiotic lists, called formularies, generally include only a few members of each antibiotic class. Formularies enable the entire hospital to take advantage of local infectious disease experts, and formularies help hospitals negotiate discount prices from pharmaceutical suppliers who want their product listed on the formulary.

Antibiotic management teams also help decide when to stop antibiotic treatment of patients who were treated without clear evidence of infection or were treated with several antibiotics at the onset of infection. As a group, physicians are reluctant to risk failing to treat an infection that would respond to treatment. Consequently, excessive prescribing occurs, particularly in intensive care units.[96] As legislative requirements increasingly mandate reporting of drug-resistant infection rates and as insurance companies increasingly refuse to pay for hospital-acquired infections, management teams will be increasingly valuable to hospitals.

Self-Medication Is Outside the Guidelines

Use of antibiotics obtained without prescription for the particular ailment is called self-medication. Antibiotics for self-medication come from over-the-counter sales at pharmacies, black market sales in local grocery stores in large cities, left-over portions of prior prescriptions, importation from countries without enforced requirements for prescriptions, and some Internet sales. A survey conducted in 2003 with 19 European countries showed that self-medication varies widely. The highest rates are in Eastern and Southern Europe where almost 20% of survey respondents acknowledged self-medication within the year covered by the survey. These rates were more than 10 times higher than those found in Northwestern Europe. In Lithuania, the self-medication rate almost equaled the prescription rate. Similar percentages were obtained when investigators went directly to homes and searched for left-over antibiotics being saved for a subsequent infection.[97,98] Pockets of Latin American immigrants in the United States are beginning to receive attention because they may have high rates of self-medication (see Box 4-5). In many Latin American countries, antibiotics are available as over-the-counter medications, a situation that creates a culture of nonchalance toward these drugs.

Box 4-5: Immigrant Self-Medication

A small survey (219 adult respondents) taken in a Latin American community of the United States probed the issue of using nonprescribed antibiotics.[99] Many respondents (45%) obtained antibiotics without a prescription when outside the United States. Moreover, a large proportion (30%) believed that a similar policy should exist in the United States. About 16% indicated that they had imported nonprescribed antibacterials into the U.S., largely for illnesses that the authors of the study thought were viral in nature. Some of the study population (24%) indicated that they would import without first consulting with a doctor. In general, respondents were more comfortable with medications from the home country, wanted to avoid visiting a doctor in the United States, and felt intimidated by a language

continues

barrier. Members of the community (19%) also purchased antibiotics without a prescription in the United States, and many (64%) felt that self-medication was preferable to going to a doctor. Part of the problem appears to be a mistrust of U.S. medicine. Another is the lack of health insurance. In this study no respondent born in the United States purchased antibiotics without a prescription. Use of black market drugs suffers from the additional problem that manufacturing standards are often lax, which reduces the effective dose and exacerbates the resistance problem.

The same general observations have emerged from other studies.[100,101] Surveys indicated that more than 30% of respondents felt that antibiotics should be available over the counter in the United States, and a quarter acknowledged using nonprescribed antibiotics in the preceding year. Small grocery stores in Latino neighborhoods of New York City routinely (34/34 surveyed) sell antibiotics over the counter.

Examples of self-medication are not limited to developing countries or immigrants. One of the most striking situations involves yeast vaginitis. Over-the-counter antifungal agents are widely available in many countries for self-treatment. Millions of women self-treat each year with these products, and if used properly, the agents are safe and effective. However, the possibility of poor compliance and consequent development of resistance looms large. The consequences may not be seen until many years later when the patient experiences immunocompromise and a life-threatening yeast infection. Persistent or recurrent infections call for consultation with a physician.

We emphasize that self-medicating puts participants at risk for emergence of resistance due to underdosing, unnecessary perturbation of the microbial ecosystem, premature cessation of treatment, and incorrect choice of antibiotic. Self-medication is also a problem for everyone because repeated cycles of antibiotic challenge, followed by pathogen outgrowth, selectively enrich resistant mutants and resistance genes that can spread from the self-medicator to the rest of the community.

Perspective

Throughout most of our antibiotic history, we have used trial and error to determine dosing regimens. The guiding principle has been to cure most patients without harm. This strategy worked so well that during the 1960s and 1970s medical microbiology began to lose the academic excitement it once had. The Surgeon General of the United States even thought that tuberculosis was controlled. Then widespread immunological changes began to occur. AIDS, cancer chemotherapy, and an aging population increased the number of persons with immune deficiencies. Moreover, increasing population density made disease transmission easier, and antibiotic consumption increased. The prevalence of resistance increased, and medical microbiology began to attract scientists again. That led to launching of new scientific journals, such as *Lancet Infectious Diseases, Emerging Infectious Diseases,* and *Microbial Drug Resistance.* Among the scientists attracted to the infectious disease problem were quantitative biologists and pharmacologists who developed correlations between animal studies and human outcome using PK/PD considerations. They reasoned that an expression for antibiotic exposure (for example, AUC_{24}/MIC) applies whether measured *in vitro* with cultured bacterial cells, with infection of animals, or with infection of humans. That permitted the antimicrobial effects of various antibiotic concentrations, and therefore doses, to be predicted for humans. The net result is a quantitative approach for empirical identification of antibiotic doses that often achieve a cure (refer to Box 4-4). However, other factors, such as host defense, are also important: Sometimes, seemingly appropriate regimens fail, and sometimes apparently inappropriate ones succeed. Even less predictable are toxic side effects. Sometimes, problems surface only after millions of persons are treated. When that happens, the offending compound is pulled from the market. The contribution of host factors and toxicity can be estimated with experimental animals,[102] but antimicrobial therapy is still far from being an exact science. We predict that empiric therapy, both physician-controlled and self-administered, will become even less reliable as subpopulations of resistant mutants expand. The relationship of our dosing strategies to these resistant subpopulations is developed in the next chapter.

Chapter 5

Emergence of Resistance

Summary: Resistance is a natural response of pathogen populations to antibiotic treatment. Changes in pathogen DNA can affect genes involved in drug uptake, drug efflux, drug inactivation, and drug-target interactions. Some changes occur spontaneously; others are induced as a response to the stress of antibiotic treatment. Still others involve the movement of whole resistance genes from one microbe to another. The enormous size of pathogen populations makes even rare events noticeable when antibiotic pressure is applied. Although many mutations are probably harmful to the organism, those that confer antibiotic resistance can be the difference between life and death for the pathogen. Consequently, they are tolerated. Subsequent mutations may then improve the fitness of mutant pathogens. The relationship between antibiotic concentration and emergence of resistance is described by the mutant selection window hypothesis. The hypothesis maintains that resistant mutant subpopulations selectively amplify when antibiotic concentrations are above MIC and below MPC (mutant prevention concentration, the MIC of the least susceptible mutant subpopulation). When we treat with antibiotics, we encourage preferential growth of resistant subpopulations by placing drug concentrations inside the selection window. Additional effects arise from mutations that increase mutation frequency. These mutations, called mutators, are enriched during long antibiotic treatment that fails to eradicate an infection. In many cases, resistance mutations are accumulated by pathogens in a stepwise fashion. However, mobile genetic elements can move several resistance genes at once. Phenotypic resistance arises without mutation and is a particular problem with some β-lactams and some bacteria.

We encourage the emergence of antibiotic resistance by using antibiotics. In this chapter, we focus on how susceptible pathogen populations become resistant. If we can severely restrict this process, we may gain enough time to develop new antibiotics before current ones succumb to resistance.

Resistance Can Emerge in Individual Patients

The existence of resistant pathogens has been known for many years. Indeed, Ehrlich and Fleming, the scientists who first discovered antibiotics, noticed resistant mutants in their laboratories. Sometimes clinical resistance appears in less than a year after a new agent is introduced into medical practice. However, when resistance is considered at the level of individual patients, it has been so uncommon that the average physician rarely observed emerging resistance. Consequently, resistance has generally been perceived as a problem for populations of people but not for individual patients; resistance has been largely an academic issue. In the early 2000s, a common refrain encountered by the authors was, "Not a problem among my patients." For many years, little incentive existed for either physicians or individual patients to make special efforts to avoid resistance. Now, however, the medical literature is replete with reports of resistance arising in individual patients. We begin this chapter with a situation involving *S. aureus* to illustrate antibiotic enrichment of resistant mutants.

In the early 2000s, *S. aureus* was recovered from a patient referred to as JH. He suffered from bacterial pneumonia, and his bacterial samples were tested for susceptibility to a variety of antibiotics. The initial *S. aureus* isolate was resistant to erythromycin, clindamycin, rifampicin, and fluoroquinolones, but it was judged susceptible to the β-lactam oxacillin, even though a relatively large fraction (0.01%) of the bacteria were resistant to the drug. For vancomycin, the MIC was 1 μg/ml, which was considered susceptible. JH was treated with both β-lactam and vancomycin. His symptoms persisted, and 2 months later another sample was taken. The MIC for oxacillin had jumped from 0.75 μg/ml to 25 μg/ml, which was clearly a sign of resistance. β-lactam therapy was stopped. During the previous 2 months, vancomycin MIC had increased from 1 μg/ml to 4 μg/ml, a value still below the resistance breakpoint. Consequently, vancomycin therapy was continued. Two weeks later, a new bacterial sample showed that vancomycin MIC had reached 6 μg/ml; after another week it climbed to 8 μg/ml. One week later JH died. In those 3 months, the pathogen population in JH lost susceptibility during treatment.[103] Comparable examples have been reported for fluoroquinolone resistance with streptococcal pneumonia,[104] and under more controlled conditions emergence of resistance is easily observed in infected laboratory animals.[105] Such data are explained by antibiotic treatment killing susceptible cells while allowing a small, resistant mutant subpopulation to grow and eventually cause resistant infection. Mutant creation and selection are described in the following sections.

Spontaneous Mutations Are Nucleotide Sequence Changes

Instructions for making all cellular components are contained in DNA molecules. (RNA serves the purpose with some viruses.) New copies of DNA are made by DNA polymerase joining nucleotides at a rate of about 800 per second. DNA polymerase occasionally makes errors. (Error rate is about 10^{-5}.)[106] Proofreading mechanisms correct some of those errors, but the proofreaders are also imperfect. (Proofreading reduces error rate to about 10^{-6} to 10^{-7}.) Another set of enzymes corrects mismatches in the DNA strands, reducing the error rate further. However, the rate is still significant in large pathogen populations. The result is that a population of progeny pathogens has a few members that differ slightly from their parents. Those differences, which at the DNA level are called mutations, are often deleterious: They lower the chance that the mutant will survive. But some mutations are beneficial within a particular environment. For example, mutations occasionally arise in a gene encoding an antibiotic target and cause changes that block binding of the antibiotic. Other mutations cause over-production of an efflux pump or lower membrane permeability. Still other mutations increase the activity of an enzyme that breaks down an antibiotic.

We can easily observe spontaneous resistance to an antibiotic by applying a large number of bacterial cells, on the order of a hundred million (10^8), to an agar plate that contains a suitable concentration of antibiotic. After incubation, a few mutant colonies appear on the plate. (Growth of the susceptible, parental cells is blocked.) The number of mutant colonies, divided by the total number of cells applied, provides an estimate of the mutation frequency. For many bacterial species and many antibiotics, that frequency ranges from one in a million (10^{-6}) to one in a hundred million (10^{-8}). We can isolate DNA from a resistant colony, and frequently we can identify the nucleotide sequence change that distinguishes the mutant from its parent. We can also place mutant DNA in the parental cell and convert that cell into a resistant mutant. Moreover, DNA recovered from resistant laboratory mutants often contains the same nucleotide sequence changes seen in resistant isolates from patients. Thus, our knowledge of resistance is quite detailed.

Mutation frequencies may seem to be small numbers, but they are large enough for some bacterial infections to contain resistant mutants prior to therapy. During pneumonia or tuberculosis, a patient may contain more than one hundred million pathogen cells. For the antituberculosis drug called isoniazid, the mutation frequency is about one in one million. Consequently, a patient with tuberculosis could contain 100 resistant mutants before therapy begins (10^8 divided by $10^6 = 10^2 = 10 \times 10 = 100$). Such calculations cause many health professionals to accept antibiotic resistance as being inevitable.

Emergence of Spontaneous Resistance Often Arises Stepwise

Two factors characterize treatment and resistance. First, infections are usually treated repeatedly. (Some antibiotics are taken four times a day because the concentration in the body drops so quickly.) Second, mutations in many different genes may lower susceptibility to an antibiotic, often with additive effects. In only a few cases is a single mutation sufficient to confer clinical resistance. (Examples are rifampicin resistance with *S. aureus* and most types of resistance with *M. tuberculosis.*) Often, multiple mutations are acquired in a stepwise fashion. In one scenario, an initial mutation arises in a gene that gives the cell a small survival or growth advantage during treatment. That mutant is gradually enriched, and in some patients it becomes the dominant member of the pathogen population. The pathogen then spreads to a new patient where the pathogen population expands before treatment is started. During this expansion phase, new (additional) mutations arise spontaneously, forming small subpopulations of double mutants. The second mutation increases the survival or growth advantage during subsequent treatment. Consequently, the pathogen with two mutations multiplies during treatment and eventually replaces the single mutant as the dominant type. This process of stepwise selection of resistance is repeated over and over as long as antibiotic pressure is present. Although these mutational events are rare, antibiotic use is very high (more than a hundred million prescriptions per year in the United States), and the number of pathogens in an infection is large, usually more than one million. Thus, the total number of pathogen cells experiencing antibiotic in a year may exceed 10 trillion.

Stepwise selection of resistance has been likened to hill climbing: A high peak is reached by gradually surmounting a series of smaller rises. Some of the clearest examples of hill climbing are seen with the fluoroquinolones. Often, drug uptake or efflux mutations are acquired first, and then a target mutation occurs in a gene encoding gyrase or topoisomerase IV. Repeated use of higher drug concentrations leads to multiple target mutations. Still other mutations affect proteins that interfere with binding of fluoroquinolones to their targets. Some clinical isolates of *S. pneumoniae* carry four topoisomerase mutations, and additional mutations can be added to these strains in the laboratory by exposing the cells to higher drug concentrations.[107] Such stepwise resistance is a logical consequence of dosing too low at the outset of treatment, an action that is taken deliberately to minimize toxic side effects or accidentally through improper self-medication.

Exceptions to the hill-climbing process, such as rifampicin resistance with *S. aureus* and *E. coli,* are equivalent to climbing a mountain with a single jump: a single resistance mutation is fully protective. In such cases, clinical resistance can arise quickly, and the antibiotic should never be used as monotherapy. When *S. aureus* colonizing the noses of patients was treated with rifampicin, almost 10% of the patients developed resistant bacteria after a few weeks of treatment.[108] Single jumps to resistance are also observed when multiple resistance genes are brought into a cell on a plasmid. This situation is discussed in Chapter 6, "Movement of Resistance Genes Among Pathogens."

Mutant Selection Window Hypothesis Describes Emergence of Spontaneous Resistance

Antibiotic-resistant mutations occur spontaneously: They arise whether we treat with antibiotic. If we do not treat, no selective pressure exists to enrich and preferentially amplify the mutants. The pathogen population will contain a tiny subpopulation of resistant mutants that will remain small. Selective pressure exists when we use enough antibiotic to interfere with the growth of the majority, susceptible population. Then the mutants have a growth advantage. The minimal amount of antibiotic needed for mutant enrichment is difficult to determine accurately, but the minimal concentration that blocks the growth of 99% of the dominant, susceptible population serves as a slight overestimate. This value is easily measured by applying a known number of bacteria to a set of agar plates containing various concentrations of antibiotic and determining the concentration that permits only 1% of the cells to grow into colonies.

For many antibiotics, high concentrations block the growth of all single-step mutants. Above such a concentration, which is called the mutant prevention concentration (MPC), a second resistance mutation must also be present for growth to occur. Consequently, fully susceptible (wild-type) cells must acquire two resistance mutations concurrently for growth above the MPC. That is expected to occur only rarely. For example, if two resistance mutations arise independently, each at a frequency of one in a million (10^{-6}), the probability that two will arise concurrently is $10^{-6} \times 10^{-6} = 10^{-12}$, one in a trillion. Keeping drug concentrations above MPC does not prevent the occurrence of mutations (changes in DNA nucleotide sequence), but it does block growth of cells after they become mutant. With respect to the hill climbing idea, keeping antibiotic concentrations above MPC forces the pathogens to jump *up* a steep cliff.

MIC and MPC define a drug concentration range within which resistant mutants are selectively enriched and amplified (see Figure 5-1; see Box 5-1 for definitions of MIC and MPC). The existence of this concentration range, called the mutant selection window,[109, 110] explains in part why a resistance problem has developed. Physicians must use doses that generate concentrations above MIC to halt pathogen growth; however, drug concentrations are kept low to minimize toxic side effects. The conventional practice of dosing to cure while avoiding side effects often places drug concentrations inside the mutant selection window for long periods of time. That enables mutant subpopulations to selectively amplify and resistance to emerge. According to the selection window hypothesis, our conventional dosing strategies contribute *directly* to the resistance problem.

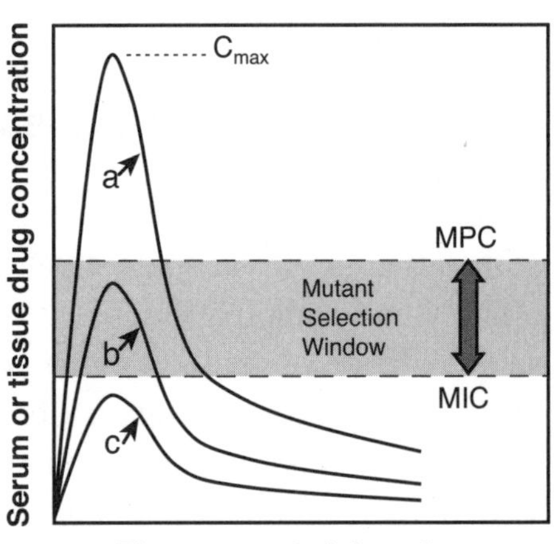

Figure 5-1 Mutant selection window. The mutant selection window, the concentration range between MIC and MPC, is shown for three hypothetical treatments that produce three hypothetical pharmacokinetic curves. Curve *a* is above the window for much of the treatment time and is expected to restrict the amplification of resistant microbial sub-populations. Curve *b* falls inside the window for much of the dosing period and is expected to enable mutant amplification. Curve *c* is below the window and exerts little selective pressure. MIC and MPC are determined from agar-plate colony-growth studies; pharmacokinetics are measured with animals or patients.

Box 5-1: Definitions of MIC and MPC

Several different designations of MIC exist. A standard procedure has been established for determination of MIC to enable comparison among clinical microbiology laboratories (refer to Box 2-1). For many bacterium-antibiotic combinations, MIC is defined as the minimal concentration that blocks growth or colony formation when 10^4 to 10^5 bacterial cells of a particular culture are tested. That value is designated as MIC without a following subscript. When the MIC of many different clinical isolates is measured, MIC_{99} is the value that exceeds the MIC for 99% of the *isolates*. (MIC_{90} is the number for 90% of the isolates.) This number is used to take into account differences among the isolates from different patients.

Because pathogen populations are heterogeneous, MIC, determined with 10^4 to 10^5 cells, need not be the same as the MIC measured with either more or fewer cells. Experimentally, the MIC for blocking the growth of 99% of the cells in a culture is usually lower than for blocking growth of 99.99% of the cells (standard MIC).[111] In such a case, the standard MIC is not the minimal concentration that blocks growth. When nonstandard values of the MIC are used, terms must be defined.

MPC is the MIC of the least susceptible, single-step mutant sub-population. With bacteria and most antibiotics, MPC can be estimated experimentally as the concentration that permits no growth (no colony) when 10^{10} (10 billion) cells are applied to drug-containing agar. Because most bacterial infections contain fewer than 10^{10} bacterial cells, drug concentrations above MPC should severely restrict bacterial population growth, including mutant subpopulation growth. Determining MPC sometimes requires large numbers of agar plates or large liquid cultures.[112, 113]

The selection window hypothesis provides a general framework for thinking about the emergence of resistance, but like all hypotheses, it is only as good as its experimental support. For example, the window should be observed even when drug concentrations fluctuate, as during oral treatment regimens. (Drug concentrations rise soon after a dose is taken and then drop gradually.) When antibiotic concentration is adjusted in bacterial cultures to mimic human treatment, mutant enrichment does occur only when fluoroquinolone concentration is inside the selection window, not when concentration is below MIC or above MPC.[114,115] The same is true in animal infections.[105] In Chapter 10, "Restricting Antibiotic Use and Optimizing Dosing," we return to the window hypothesis when we discuss how to adjust dosing to restrict emergence of resistance.

Mutations Can Be Caused (Induced) by Antibiotic Treatment

Bacteria, and probably other microbes, have ways to increase their mutation rates when placed under stress. One of the best-understood mechanisms is the bacterial SOS response. In this system, the expression of more than 30 genes is controlled by a protein called LexA, which binds in front of each of the SOS genes and blocks formation of mRNA, thereby blocking formation of proteins encoded by the genes. Under conditions of stress, such as DNA damage caused by ultraviolet light or fluoroquinolone treatment, the LexA protein cleaves itself. The resulting loss of LexA permits proteins to be made from the information in the SOS genes. Some of these proteins participate in DNA damage repair that involves synthesis of new stretches of DNA. Those new stretches may not perfectly match the original nucleotide sequence, thereby generating mutations. Consequently, SOS-dependent repair is often called error-prone repair. After the stress is removed, the signal that led to degradation of LexA dissipates; newly made LexA protein binds upstream of the SOS genes, and their expression is silenced. That resets the system for a subsequent response to stress. However, the mutations generated during induction of the SOS response remain. If they confer a growth advantage, the cells containing them will overgrow the parental, nonmutant cells.

Induced mutants, such as those created by the SOS response, are observed by applying a bacterial culture to agar plates containing a fluoroquinolone. As the plates are incubated, colonies arise over time.[116] Those seen after a day or so represent resistant mutants present spontaneously in the population prior to drug exposure. Over the next week, colonies gradually arise, but not if the strain contains a *lexA* mutation that prevents induction of the SOS response. Conversely, induction of mutants occurs to a higher level if a mutator mutation (discussed below) is present.[117] Induced resistance is blocked by antibiotic concentrations high enough to kill susceptible cells, because the susceptible population is the source of new mutants.[82] Thus, keeping doses high is important for preventing the induction of resistance and for preventing selective growth of preexisting mutants.

Resistance Arises from Several Molecular Mechanisms

To be effective, most antibiotics must get inside pathogen cells. Microbial pathogens are surrounded by a membrane and in some cases by a tough cell wall. Neither covering is impenetrable: A variety of molecules go in and out of cells. Changes in either membrane or wall can reduce the uptake of an antibiotic, thereby reducing its intracellular concentration and effectiveness. Such changes give rise to what are often called low-level resistance mutants because the resulting decrease in susceptibility is often not great.

Many microbes also contain proteins that pump out noxious substances, including antibiotics. The bacterium *Pseudomonas aeruginosa* is particularly well protected by efflux pumps (see Box 5-2), as are some species of the yeast *Candida*. Increased pumping and the associated loss of antibiotic susceptibility can arise through mutations that raise the synthesis of pump components. Some pumps remove drugs of several types (multidrug pumps). With these pumps a single mutation reduces susceptibility to several classes of agent simultaneously. Such pumps are a major reason why we worry about excessive use of disinfectants and antiseptics. These agents permit efflux mutants to selectively amplify and lower susceptibility to several antibiotics at once.

Box 5-2: Efflux Pumps of Pseudomonas

P. aeruginosa, a common inhabitant of the environment, can cause serious pneumonia. The microbe has at least 10 efflux systems. The four most extensively studied are called MexAB-OprM, MexCD-OprJ, MexEF-OprN, and MexXY-OprM.[118] Some systems, such as MexCD-OprJ, are normally turned off, at least during laboratory cultivation of the microbe. (No protein is expressed from the genes.) When *P. aeruginosa* is treated with disinfectants or antiseptics, such as alcohol or chlorhexidine, the pump is made by the bacterium, thereby lowering bacterial susceptibility to the agents. Mutations that eliminate the off switch for the pump genes raise MIC.[119] Examination of clinical isolates having reduced susceptibility to ciprofloxacin reveals that some (3%) have mutations affecting MexCD-OprJ. These isolates tend to be found in patients who have been treated with ciprofloxacin for long times.[120]

A third molecular strategy for resistance is production of proteins that chemically destroy the antibiotic. Among the most notorious examples are β-lactamases (see Box 5-3). These enzymes break down members of the antibiotic class that includes penicillin. The genes encoding β-lactamases often move at high frequency from one bacterium to another as parts of plasmids, which has made β-lactamase production one of the most important resistance problems. The pharmaceutical industry created β-lactams that are not cut by the common β-lactamases, but new β-lactamases evolved that attack the new β-lactams. Drug companies responded with β-lactamase inhibitors that were combined with β-lactams to provide an effective therapy. The bacteria continued to evolve, acquiring mutations that rendered the β-lactamases insensitive to the inhibitors. This scenario illustrates how keeping one step ahead of bacteria is not enough—we need to stay two or more steps ahead.

Box 5-3: β-lactamases and Clavulanic Acid

β-lactams, such as penicillin, are naturally occurring antibiotics. Many bacteria (both Gram-positive and Gram-negative) have enzymes called β-lactamases that break down β-lactams, thereby providing a defense for the bacteria (hundreds of different β-lactamases have been identified).[121] The β-lactamases are divided into four classes (A, B, C, and D). The A family includes a group of plasmid-encoded enzymes called TEM. (TEM is derived from the name of the *E. coli* strain in which the β-lactamase was found.[122]) The TEM enzyme, which is common among *E. coli* strains, breaks down penicillin but not ceftazidime, a third-generation cephalosporin. A single amino acid change in TEM creates TEM-12, and an additional change produces TEM-10 or TEM-26, depending on the particular change. These mutant enzymes readily degrade ceftazidime and are called extended-spectrum β-lactamases (ESBLs). Outbreaks of ESBL-producing *Klebsiella pneumoniae* have created serious problems that require hospitals to switch from use of ceftazidime to other antibiotics.

Some species of the soil microbe *Streptomyces* produce weak β-lactams that act as inhibitors of the β-lactamases. One of these is called clavulanic acid. When purified clavulanic acid is added to amoxicillin, a β-lactam, it lowers the MIC with some β-lactamase-producing strains of *S. aureus* from 500 μg/ml to 0.1 μg/ml. Thus, clavulanic acid is commonly added to β-lactam treatments. In 2008, clavulanic acid had sales of more than one billion dollars.[123] Other commonly used β-lactamase inhibitors are sulbactam and tazobactam.

A fourth resistance mechanism is illustrated by the protein targets of antibiotics. They change, thereby blocking the binding of the antibiotic. The targets of fluoroquinolones acquire amino acid changes in a specific region thought to be part of the drug binding site (see Box 5-4). Protein targets can also increase in number, thereby requiring more antibiotic to kill the bacteria. Such is the case with β-lactamases.[124] Target mutations do not always affect proteins: For some inhibitors of protein synthesis, resistance is due to changes in the nucleotide sequence of ribosomal RNA. One form of resistance to erythromycin falls in this category.

Box 5-4: Fluoroquinolone-Resistant Gyrase Mutants

For target-based fluoroquinolone resistance, mutations cause amino acid changes in a section of gyrase and topoisomerase IV called the quinolone-resistance-determining region. The protein targets are quite similar in many different bacteria; consequently, protective amino acid changes are similar for a wide variety of bacteria. The two most common changes occur at amino acid numbers 83 and 87 in the *E. coli* numbering system for the gyrase A protein. Tests have been established using DNA to identify resistant isolates from patient samples. Such analyses are expected to be quite useful with *M. tuberculosis* because a DNA test can be completed in less than a day, whereas colony testing requires 3 weeks. With susceptible (wild-type) *M. tuberculosis*, the amino acid normally at position 83 is a type associated with resistance in many other bacteria. Thus, *M. tuberculosis* naturally has lowered susceptibility to fluoroquinolones.

A fifth type of resistance mechanism is observed when a protein other than the drug target interferes with drug binding. An example involving fluoroquinolones and Qnr is described in Box 5-5.

Treatment Time Can Contribute to Resistance

If drug concentration drops below the MIC before infection is cleared, residual pathogens may regrow and cause disease relapse. During regrowth, spontaneous errors in DNA replication generate new resistant mutants, and the resulting mutant subpopulations will be enriched when antibiotic therapy is restarted. Successive rounds of treatment and relapse with the same type of antibiotic contribute to the emergence of resistant pathogen populations. Traditional

Box 5-5: Fluoroquinolone Resistance Due to a DNA Mimic

The fluoroquinolones trap an enzyme called gyrase on DNA. With purified enzyme, the drugs only work if DNA is also present, indicating that the target of the drug is a DNA-gyrase complex, not the enzyme alone. A protein called MfpA interferes with the binding of fluoroquinolones to gyrase-DNA complexes. MfpA has a structure similar to DNA, which causes gyrase to bind to the protein rather than to DNA. When MfpA is present, the fluoroquinolone cannot make the proper attachment to gyrase.[125] A gene called *qnr* encodes a protein, Qnr, that is similar to MfpA. The *qnr* gene is now found on plasmids in many parts of the world and contributes to fluoroquinolone resistance. The natural function of Qnr and MfpA is not known.

treatment strategy calls for antibiotic therapy to be long enough to eliminate most, if not all, of the susceptible pathogen population. Whether the usual 10-day period is long enough or too long is not known. Longer treatment places more selective pressure on commensal bacteria, and in principle it contributes to enrichment of resistance genes that spread through the ecosystem. Careful clinical trials are now needed to determine how long treatment should be to restrict the emergence of resistance.

Mutator Mutations Increase Mutation Frequency

Some bacteria acquire changes in the proteins that make DNA or repair errors in DNA. Those alterations raise the frequency for the occurrence of additional mutations, sometimes by a thousand fold. Because these mutator mutations tend to be harmful to pathogens, mutator mutants are usually rare members of a bacterial population. However, when antibiotic pressure is applied, mutators may increase their relative abundance.[126]

Mutators are clinically obvious in cases of cystic fibrosis, a genetic disease of humans in which a symptom is the formation of thick mucous in the lungs. *Pseudomonas aeruginosa* grows well there and is difficult to kill with antibiotics. Cystic fibrosis patients receive repeated antibiotic treatments, and as a result, *P. aeruginosa* mutators are selectively enriched.[127] That causes *P. aeruginosa* to become antibiotic resistant at high frequency, making it extremely difficult to eradicate.

Phenotypic Resistance Occurs Without Mutations

Resistance can arise without changes in DNA. The stress of antibiotic treatment causes cells to produce protective proteins using existing genes. An example is induced expression of β-lactamase genes by some Gram-negative bacteria. Release of protective enzymes into the growth medium then enables neighboring cells to grow in the presence of the antibiotic. As pointed out in Box 5-3, many β-lactamase genes exist. Some are inducible, whereas many others are always active.

Resistance May Compromise Antiseptic and Disinfectant Use

An increasing public awareness of pathogens, particularly antibiotic-resistant ones, has stimulated the antiseptic and disinfectant industry to encourage consumers to "clear your home of dangerous germs." This movement has at least two negative features. First, we may be doing our children a disservice by reducing their exposure to microbes: Exposure to "dirt" may be important for proper immune system development.[128] Second, resistance to antiseptics and disinfectants may remove our last line of defense against pathogens on our skin and on environmental surfaces; it could seriously compromise surgical procedures.[129] Manufacturers of disinfectants now need to examine large numbers of households to determine whether resistant mutants in the environment increase following disinfectant use.

Viral Resistance Can Arise Readily

Viruses cause their host cells to make new virus particles. In some cases, the host cell breaks when it is full of virus; in other cases, the viruses bud out through the cell membrane. In either case, large numbers of virus particles may be produced. Extensive copying of viral genomes provides many chances for mutations to arise, especially because viral RNA and DNA polymerases are often not accurate. (They tend to lack proofreading activities.) Indeed, so many errors are introduced during replication that some RNA viruses are considered to be quasispecies. Persons infected with HIV are thought to harbor a swarm of viruses, sometimes containing over one billion members of many different types. When an antiviral agent is applied, mutant virus readily emerges due to its selective advantage. Large population sizes and inaccurate polymerases make resistance such a serious problem with HIV that chemotherapy generally involves three or more drugs at once. Indeed, resistance dominates the thinking behind HIV treatment (see Box 5-6).

Box 5-6: HIV Treatment and Resistance

HIV-1 is considered to be a quasispecies in every patient.[130] This means that many genetically related subpopulations comprise the large viral population, which replicates frequently and inaccurately. Some members of the population are likely to have two resistance mutations before therapy begins, which permits enrichment of resistant mutants when only two drugs are used. Because there are now six different classes of antiHIV drug (see Table 3-1), combinations can be applied. Each compound has a distinct genetic barrier that affects its utility. For example, some agents require only a single mutation for high-level resistance, whereas others can be dosed at high enough concentrations to require several viral mutations for growth in the presence of drug.[66] Many of the nucleoside reverse transcriptase inhibitors (NRTIs), the non-nucleoside reverse transcriptase inhibitors (NNRTIs), and the fusion inhibitors require only a single mutation. Integrase inhibitors and second-generation NRTIs and NNRTIs may require two; compounds such as zidovudine, an NRTI, may require several.[66] The compounds also differ in their antiviral effectiveness: A compound that more severely reduces the viral load is expected to be less likely to permit evolution of new resistant mutants during therapy.

It is thought that prior to treatment few members of the viral population have mutations to three different drug classes. It is reasoned that using a combination therapy with three drugs can keep the patient at least one mutational step ahead of the virus, depending on the particular drugs chosen. When the viral population catches up, the regimen is changed. Experience with bacterial pathogens indicates that staying only one step ahead of the pathogen is insufficient to restrict the emergence of resistance. Thus, the initial choice and number of compounds used for HIV therapy is important for the long-term success of treatment.[131]

Resistant mutants do evolve, and they can dominate the population. When they spread to other persons, therapy options are greatly reduced. Thus, primary resistance, which is resistance obtained from other infected individuals, is an important factor. (Examples are listed in Table 5-1.) To manage the problem of transmitted resistance, patient

continues

samples are tested for resistant mutants before treatment. Testing usually relies on determining viral nucleotide sequences. Interpretation can be difficult, because the importance of any given mutation to resistance may not be well documented. Moreover, complex effects can arise from the presence of multiple mutations, both inside and outside the target gene. We stress that only majority species are usually detected,[132] which leaves many subpopulations that can be enriched by antiviral treatment. The complexity is emphasized by virologic failure sometimes lacking a clear association with a particular mutation.[132]

Because resistant mutants of HIV frequently suffer a fitness disadvantage, removal of antibiotic pressure enables the wild-type version to regain dominance.[131,132] However, this reversion is a slow process and usually incomplete, which results in mixtures of wild-type and mutant virus persisting for many years.[133]

Table 5-1 Prevalence of Resistance to AntiHIV-1 Agents: USA 2003[134]

Drug Class	Prevalence of Resistance*
One or more drug	10%
Nucleoside reverse transcriptase inhibitor	3%
Non- nucleoside reverse transcriptase inhibitor	6%
Protease inhibitor	2%

* Samples were from 317 patients from 40 cities across the United States. Numbers represent the presence of mutations associated with resistance.

Resistance Mutations Can Affect Pathogen Fitness

Accumulation of mutations can interfere with pathogen growth, that is, the mutations reduce fitness. In principle, reduced fitness could permit susceptible members of the population to overgrow the mutants when treatment with a particular antibiotic is halted. (Such a situation is often seen with HIV-1[131,132] and with several bacterial pathogens thought to acquire resistance through the use of animal growth promoters.[135]) After a suitable wait, treatment with the antibiotic could be restarted. We do not know how long to withhold an antibiotic, because mutant subpopulations can remain at elevated levels, thereby facilitating rapid regrowth of resistant mutants when infections are re-exposed to the antibiotic. Moreover, loss of fitness can be corrected by yet another mutation, thereby enabling the mutant pathogen to regain fitness and also remain resistant. Cycling antibiotics has not solved the resistance problem.

Unintended Damage Can Arise from Treatment

Our bodies are veritable ecosystems for microbes (refer to Box 4-1), and some of the commensal organisms keep pathogen populations from expanding to harmful levels. Broad-spectrum antibiotics kill beneficial organisms that normally limit the growth of harmful ones. Many digestive and vaginal problems arising from use of broad-spectrum antibacterials are correctable by replacing the normal flora after treatment. This often occurs naturally. However, serious diarrhea can arise from the bacterium *Clostridium difficile*.

C. difficile is a rod-shaped bacterium (see Figure 5-2) that forms spores, tough structures that permit the organisms to survive harsh conditions such as cooking. *C. difficile*, which is a common inhabitant of animal and human digestive tracts, is most problematic for persons admitted to a hospital or long-term care facility and then treated with a broad-spectrum antibiotic, such as a fluoroquinolone or cephalosporin (see Box 5-7). Loss of the normal bacteria in the digestive tract facilitates growth of *C. difficile* and toxin production; severe diarrhea and sometimes death follow. *C. difficile* is thought to be acquired from the hospital environment as human-to-human transmission.[136] The organism is also found in a variety of farm animals. (Horses and young pigs can experience life-threatening disease following antibiotic treatment.) A 2006 survey of retail meat in Canada revealed that 6% of the samples were contaminated by *C.*

Figure 5-2 *Clostridium difficile.* Scanning electron micrograph of *C. difficile* obtained from a stool sample at a magnification of 2,905 times.

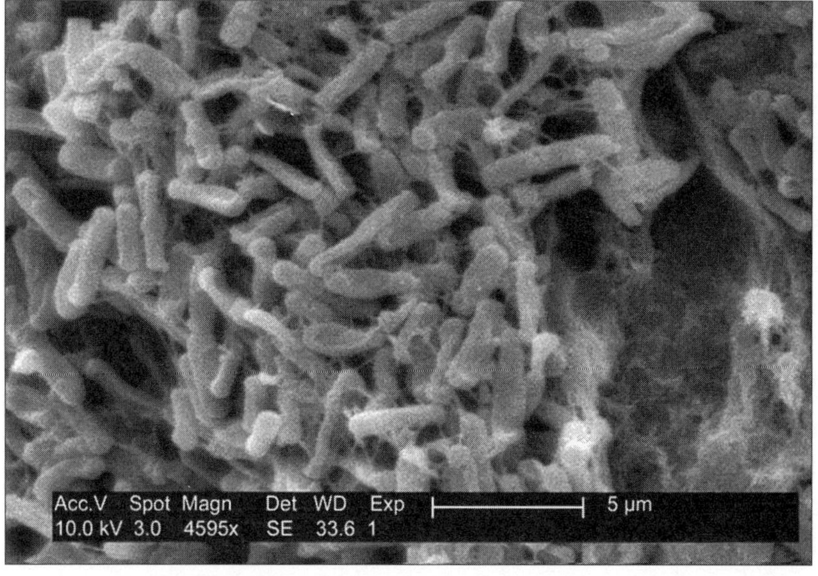

Public Health Image Library # 6258; photo credit, Janice Carr.

Box 5-7: Clostridium difficile

As our antibiotics have become increasingly potent, diarrhea associated with *C. difficile* emerged.[139,140] In the United States, diagnosis of this disease doubled from 2000 to 2003, mainly among patients above age 65 in short-stay hospitals.[141] The colonization frequency, 10%–25% among hospitalized patients, is only 2% to 3% in the general population.[142] In 2005, the bacterium was the leading cause of hospital-associated diarrhea, and it caused more deaths than all other intestinal infections combined. (Its dramatic increase is shown in Figure 5-3). In 2006, a fluoroquinolone-resistant outbreak strain, producing increased levels of toxins, spread among hospitals; by 2007, *C. difficile* was responsible for raising U.S. healthcare costs by 3.5 billion dollars per year.[143] *C. difficile* is a spore-former: consequently, alcohol-based antiseptics may not be effective. (Spores are not easily killed by alcohol.) Indeed, part of the spread of *C. difficile* has been blamed on the widespread use of alcohol-based hand sanitizers that had replaced soap and water.[141]

difficile;[137] in Tucson, Arizona 40% of the meat samples were contaminated in a 2007 study.[136] The organism has also been detected in ready-to-eat packaged salads in Scotland.[138] Thus, *C. difficile* is widely distributed.

Another example of unintended damage appears to have arisen from the extensive use of chloroquine for treatment of malaria. Chloroquine and the antibacterial fluoroquinolones are structurally related; consequently, the possibility of cross-resistance has long been thought possible. A recent report[144] describing a study in a remote region of Guyana identified an isolated human population that had access to chloroquine but not to fluoroquinolones. Examination of the study population revealed that almost 5% carried ciprofloxacin-resistant *E. coli* shortly after an epidemic of malaria had been treated with chloroquine. (As a point of reference, ciprofloxacin-resistance in a U.S. critical care unit at the time was 4%.) DNA-based analysis of the *E. coli* isolates showed that many different strains were present in the region of Guyana, arguing for independent emergence of resistance in the study population and against a rare visitor from the outside world having introduced a resistant strain into the study region. If chloroquine treatment of malaria drives the emergence of fluoroquinolone resistance in bacteria, many regions of the Earth will be affected.

Figure 5-3 Increasing mortality due to *C. difficile* in the United States.

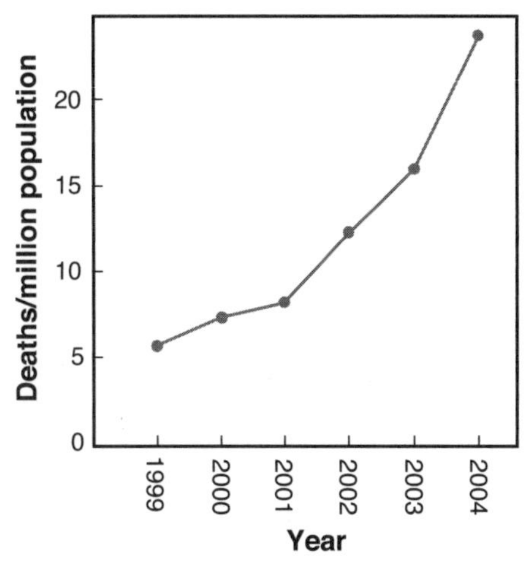

Source: Redelings, M., Sorvillo, F., Mascola, L. "Increase in *Clostridium difficile*-Related Mortality Rates, United States, 1999–2004." *Emerging Infectious Diseases* 2007; 13:1417–1419. Similar data have been reported for other countries, including Singapore and Finland. Sources: Lim, P., Barkham, T., Ling, L., Dimatatac, F., Alfred, T., Ang, B. "Increasing Incidence of *Clostridium difficile*-Associated Disease, Singapore." *Emerging Infectious Diseases* 2008; 14:1487–1489 and Lyytikäinen, O., Turunen, H., Sund, R., et al. "Hospitalizations and Deaths Associated with *Clostridium difficile* Infection, Finland, 1996–2004." *Emerging Infectious Diseases* 2009; 15:761–765.

Perspective

Several aspects of pathogen-antibiotic relationships favor the emergence of resistance. The huge size of pathogen populations is one. A single infection can contain 100 million bacteria, and the number of pathogen-infected persons and animals is large. For example, in the United States we issue more than 100 million prescriptions per year to humans. The number of food animals treated may be even higher; although accurate numbers are not available, animals raised for food receive 10 times more antibiotic tonnage than humans. (Antibiotic use through veterinary care of individual animals is probably less than human antibiotic consumption.) Consequently, subpopulations of resistant mutants can constitute only a tiny fraction of the total pathogen population and still be a large number. A second aspect is our collective attitude. Over the last 6 decades, the primary objective of antibiotic treatment has been to cure disease with few side effects. We have not tried hard to stop resistance. Third, societies continue to encourage antibiotic use, even though clinical research has established that increased use parallels increased resistance. For example, pharmaceutical companies make more money by increasing sales (use), and in some countries physicians make much of their income from writing prescriptions. Farmers claim that without antibiotics they cannot make a profit. A fourth factor is the movement of resistance genes among bacterial species, sometimes as groups of genes that confer resistance to multiple antibiotics. This horizontal spread of genes is the focus of the next chapter. These four factors—huge numbers of pathogens, dosing to cure, encouraging antibiotic use, and spread of resistance genes among bacterial populations—work together to drive the emergence of resistance.

Chapter 6

Movement of Resistance Genes Among Pathogens

Summary: Resistance genes move from one microbe to another by several mechanisms. These processes, which are best understood with bacteria, operate through plasmids (conjugation), viruses (transduction), and in some cases direct uptake of DNA from the environment (transformation). Bacteria contain genetic elements called transposons that move genes from one DNA molecule to another. Consequently, chromosomal genes that acquire resistance mutations through spontaneous events can be mobilized by being moved to plasmids that then transfer to other bacteria. Bacterial cells also contain DNA elements called integrons that can assemble resistance genes into short regions of a chromosome. Those regions can move to plasmids and then to other bacteria. Movement of resistance genes involves specific nucleotide sequences at which specific proteins act. Those proteins are potentially subject to man-made inhibitors.

In previous chapters, we emphasized resistance that is passed vertically, from mother cell to daughter cell. We now turn to the movement of genes from one microbial cell to another, a process that is called horizontal transfer. This phenomenon can occur at a much higher frequency than spontaneous mutation, and it is causing the resistance problem to grow at an increasing rate. We can do almost nothing about horizontal gene transfer; consequently, much of the discussion is still at the level of basic biology. We begin with an overview.

Horizontal Gene Transfer Involves Specific Molecular Events

Horizontal gene transfer mechanisms can move many genes in a single transfer event as regions of DNA travel from one bacterial cell to another. Three intercellular mechanisms are known: conjugation, transduction, and transformation. Conjugation occurs when self-replicating DNA molecules called plasmids move sets of genes among bacteria that are not necessarily closely related. The second mechanism, transduction, is mediated by bacterial viruses (bacteriophages). During transduction, bacterial DNA is incorporated into phage particles and then transferred to a new bacterial cell during infection. With transformation, the third type of transfer, DNA released from one cell is taken up by a nearby, recipient cell.

Two intracellular processes move sets of genes from one DNA location to another and from one DNA to another. In the process called transposition, a discrete region, the transposon, moves, often after being replicated. The second process, cassette integration, assembles tandem arrays of genes and provides for their expression (formation of mRNA and protein from the information in the genes). Each process is sketched below after a brief description of genetic recombination and plasmids.

Recombination Involves Breaking and Rejoining of DNA Molecules

A crucial step in many types of gene movement is the breaking and rejoining of DNA molecules. When DNA segments are rearranged, the process is called genetic recombination. Two types occur. With homologous recombination, also called general recombination, exchange of nucleotide sequence information occurs between two DNA molecules that have similar nucleotide sequences (see Figure 6-1). This process often begins with a double-strand break in DNA. Single-strand degradation from the ends exposes single-stranded regions, one of which invades a nearby DNA duplex having a similar nucleotide sequence. The invading strand displaces one strand of the recipient DNA, forming complementary base pairs between the invading and recipient DNA, usually over a short region. The displaced strand of the recipient DNA then forms complementary base pairs with the noninvading single strand of the first DNA. Breakage of the DNA strands at the junction points, followed by sealing of breaks, results in an exchange of information between two DNA molecules. Proteins involved in homologous recombination facilitate the alignment of single strands, breakage of DNA, and resealing of DNA breaks.

Homologous recombination enables sections of DNA from the environment to be incorporated into microbial chromosomes if the resident and incoming DNA share regions of nucleotide sequence similarity. Homologous recombination also permits mutations, such as those responsible for antibiotic resistance, to move from the chromosome to a plasmid and vice versa, providing that the two DNA molecules contain similar sequences for strand invasion and complementary base pairing.

Site-specific recombination is a second type of DNA rearrangement. It occurs at sites on DNA that lack a high degree of nucleotide sequence similarity. An example is observed when a viral DNA inserts into chromosomal DNA of the host bacterial cell. Recipient host DNA is broken at a specific location recognized by a viral protein (integrase), and ends of the inserting

Figure 6-1
Homologous recombination. **A**. DNA crossover through a process of DNA strand breakage and rejoining. In the simplified example, a double-strand break in DNA #1 leads to partial digestion and single-strand ends. These ends invade DNA #2 and form complementary base pairs. Breaks in DNA #2 (open triangles) and sealing of breaks lead to two recombined DNA molecules. **B**. Double crossover leads to exchange of a region of DNA. Two DNA molecules pair such that crossing over occurs in two regions (dotted circles), as shown in part A. That enables regions to be exchanged.

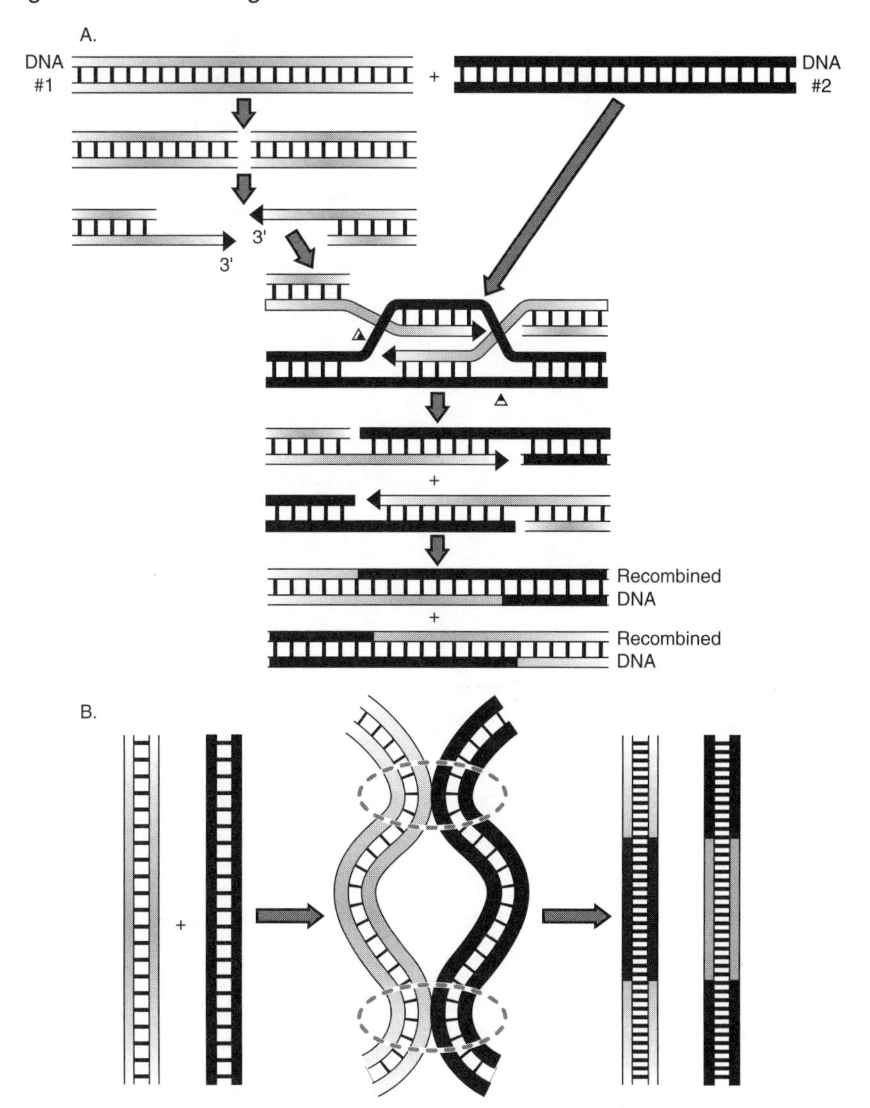

DNA, also recognized by integrase, join to the broken ends of the host DNA (see Figure 6-2). The reverse reaction (excision) also occurs.

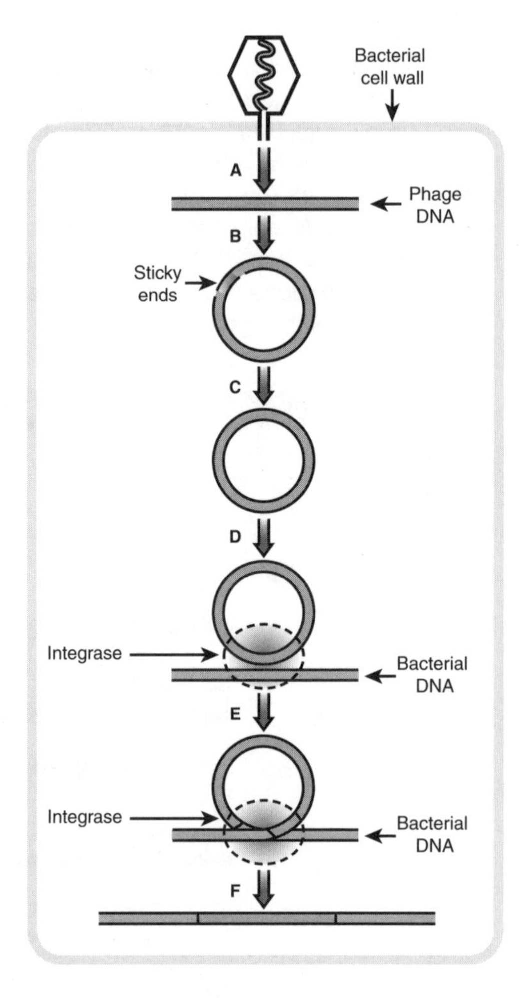

Figure 6-2 Integration is the process in which one DNA molecule inserts into another. In the example, a bacteriophage is shown injecting its DNA into a bacterial cell (*A*). The DNA circularizes (*B*), and the sticky ends (short, complementary regions of single-stranded DNA at the ends) are ligated (*C*). The phage DNA and the bacterial chromosome are brought together at specific sites by a protein called integrase (*D*). The two DNA molecules break and rejoin (*E*) to integrate the phage DNA into the bacterial chromosome (*F*).

Plasmids Are Molecular Parasites

Plasmids are autonomous DNA molecules that are molecular parasites. They are the major DNA vehicles that move antibiotic resistance genes among bacteria. Each plasmid contains its own origin of replication and a gene encoding a specific replicator protein. These two elements give plasmids control over their own DNA replication, over their own "lives." Plasmids can be small (on the order of 3,000 base pairs) or large (chromosome-size or about 3,000,000 base pairs). Many are circular, but linear forms are found. Some plasmids are present at only one or a few copies per cell, whereas others maintain hundreds of copies. The low-copy-number plasmids have special mechanisms for assuring that after cell division each daughter cell receives a plasmid copy. One mechanism, based on addiction modules, is briefly discussed in Box 6-1. Many plasmids contain a variety of antibiotic-resistance genes, which enables them to transfer several types of resistance at once. Plasmids create serious resistance problems because they can move from cell to cell at high frequency.

Box 6-1: Addiction Modules

Some plasmids ensure that all cells in the population contain a plasmid by killing any cell that fails to get a plasmid copy during cell division. These plasmids contain a pair of adjacent genes called an addiction module. One of the genes encodes a potent toxin protein capable of killing bacterial cells. The toxin protein is stable, that is, it is not easily degraded by other proteins (proteases) that normally break proteins. The second gene encodes an antitoxin, a protein that binds tightly to the toxin or the toxin's target, keeping the toxin from killing the cell. The antitoxin is easily broken. As long as the plasmid is present to make antitoxin, the cell is safe. But if a cell divides and one of the daughter cells fails to get a copy of the plasmid, that cell cannot make new antitoxin or toxin. (No gene is present for their production.) Soon after cell division, the existing antitoxin in the plasmid-free cell breaks down, permitting the more stable toxin to kill the cell. Thus, only plasmid-containing cells live.

Many bacteria contain multiple toxin-antitoxin genes in their chromosomes. Their functions are poorly understood, but it is becoming apparent that some contribute to protecting cells from stress by degrading mRNA, thereby blocking protein synthesis and shifting the cell into a dormant state. Artificial manipulation of the toxins and their intracellular targets may be a way to improve the lethal action of existing antibiotics.

Some Plasmids Move by Conjugation

Some bacterial plasmids encode a mating apparatus that enables them to move from one cell to another. The process, called conjugation, is the principal way in which antibiotic resistance genes and virulence factors move among bacteria. During conjugation a "male" plasmid-containing cell binds to a plasmid-free "female" cell. Plasmid DNA in the male is copied, and a copy of the male DNA is transferred to the female cell. Conjugation can be observed in a simple way. If a male cell is resistant to one antibiotic and a female cell is resistant to another, conjugation can be detected by placing a drop of male culture and a drop of female culture together on an agar plate containing both drugs. The two drops mix, the two cell types come together, and male DNA moves into female cells. At about the same time, the plates dry. When the plates are incubated overnight at body temperature, colonies resistant to both drugs appear at high frequency on the plates where the culture drops were placed. Because neither the male cells nor the female cells can grow on the agar, genes from one cell type (by definition the male) must have passed into the other (the female type), thereby creating cells that are resistant to both drugs.

Conjugation requires that the plasmid have genes responsible for the transfer process. The best-studied system is the fertility factor (F-plasmid) of *E. coli.* This large (100,000 base-pair), conjugative plasmid contains a set of genes called *tra* whose protein products form hair-like appendages (pili) on the bacterial surface. Pili are thought to be important for bringing male and female cells close enough for DNA transfer to occur.

Mutations in genes that confer antibiotic resistance can move from one bacterial cell to another by conjugation if the genes are part of a conjugative plasmid. When conjugation is followed by recombination, the resistance mutations can move to the bacterial chromosome of the recipient cell. Sometimes the entire plasmid inserts into the host chromosome. Subsequent conjugation then causes a copy of the host chromosome to be pulled into the female cell along with the plasmid. Sometimes the whole chromosome moves; sometimes it breaks before the process is complete.

Bacteriophages Move Bacterial Genes by Transduction

Viruses that attack bacteria are called bacteriophages. They are found in a variety of sizes and shapes. Some are long filaments, some are almost spheres, and still others look like miniature hypodermic syringes (see Figure 6-3). The latter attach to the surface of bacteria and "squirt" their DNA into the host cell. Intracellular phage DNA is then transcribed into mRNA, which is translated into viral proteins. Some of these proteins replicate phage DNA, whereas others form viral parts. When the necessary phage components have been produced, new virus particles assemble spontaneously. In many cases, host cells break apart, releasing phage to infect other cells. Such phages are called lytic phages because they lyse their host cells. These phages constitute a potential treatment for bacterial infections (see Box 6-2).

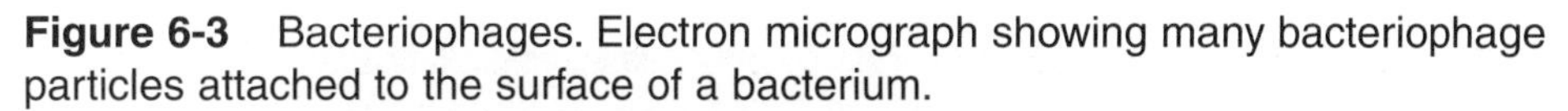

Figure 6-3 Bacteriophages. Electron micrograph showing many bacteriophage particles attached to the surface of a bacterium.

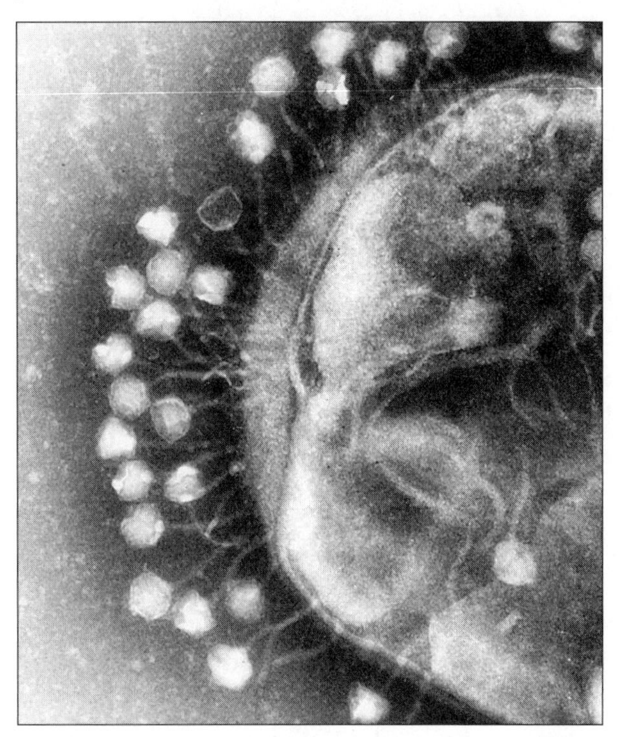

Source: http://en.wikipedia.org/wiki/File:Phage.jpg; author: Graham Colm.

Box 6-2: Phages as Therapeutics

Bacteriophages are currently being considered as a way to control bacterial infections without antibiotics. Phages tend to be narrow-spectrum agents, sometimes infecting only a single bacterial species. Consequently, some of the issues of antibiotic usage associated with broad-spectrum agents do not apply. However, bacterial mutations that confer resistance to phages have been known since the early 1940s.

One of the problems with phage therapy is that phage-resistant mutations tend to be highly protective. Consequently, the equivalent of MPC cannot be achieved by increasing phage concentration; combination therapy with two different phages that lack cross resistance needs to be applied. The scientific literature is beginning to document efforts to use phage against *Salmonella* in chickens, for coral white-plague disease, and nasal colonization by MRSA. Mice infected with the bacterial pathogen *Klebsiella* were saved from death by

continues

treatment with phage at the time of infection. However, a delay of phage treatment by 6 hours rendered the phage ineffective.[145] Much more work is required before phages become a viable alternative or addition to antibiotics.

When progeny virus is made, a piece of bacterial DNA is occasionally incorporated by mistake into a phage particle. In this situation, the bacterial genes replace viral genes. The virus can still inject DNA, but bacterial rather than viral genes enter the new host bacterium. In this process, called generalized transduction, the phage transfers bacterial genes from one bacterium to another; bacterial genes are then incorporated into the bacterial chromosome by homologous recombination.

Some phages have the ability to insert (integrate) their DNA into the host chromosome. The phage DNA encodes a protein (repressor) that blocks the production of other phage proteins, thereby silencing the phage genes. This process of integration and gene silencing is called lysogeny, and phages that carry it out are called lysogenic bacteriophages. Certain types of stress, such as transient treatment with fluoroquinolone or ultraviolet light, cause the phage repressor to break, thereby permitting expression of the phage genes. That leads to excision of viral DNA, formation of viral parts, and lysis of host cells. When the DNA of a lysogenic phage excises from the host chromosome, it occasionally cuts out adjacent bacterial DNA. That chimeric DNA can be incorporated into a progeny virus particle. When the phage with chimeric DNA infects a new host, bacterial DNA is carried with phage DNA. Subsequent phage integration causes the bacterial genes carried with the phage DNA to become a part of the chromosome of the new host bacterium. This process is called specialized transduction because only pieces of DNA near the phage integration site are moved.

Bacterial Transformation Involves Uptake of DNA from the Environment

DNA molecules can also be taken up by bacterial cells directly from the environment. Transformation requires no special gene in the transferred

DNA. However, in some bacteria, such as *Bacillus subtilis* and *Haemophilus influenzae*, a set of genes in the recipient cells facilitates transformation. In *B. subtilis*, these genes are expressed when a bacterial culture stops growing exponentially. Cells that readily take up DNA are said to be "competent." In *B. subtilis*, incoming DNA is broken and rendered single-stranded, which then facilitates integration into the bacterial chromosome via homologous recombination.

Transposition Moves Genes from One DNA to Another

Transposons are mobile DNA elements that reside in DNA molecules. They have discrete ends that help define and identify each element, and they usually carry one or more gene that helps them move to other DNA molecules. (To move from one DNA to another, the donor and recipient DNA must break and rejoin; specific proteins are involved in this process.) When transposons contain an antibiotic resistance gene, they can be easily followed because they cause bacterial cells to be drug resistant. Transposons enable resistance genes to hop from chromosome to plasmid and vice versa. When transposons hop into each other, they can create multidrug-resistant mobile DNA elements.

Gene Mobilization Moves Genes from the Chromosome to a Plasmid

Mobilization has been studied extensively with genes encoding proteins that break down β-lactams, such as penicillin. After mobilization, the genes encoding β-lactamases continue to evolve. For some types, it has been possible to identify hundreds of varieties that originally came from one or a few mobilization events. Consequently, good evidence exists for mobilization leading to gene movement among species. Gene mobilization is part of the reason that indiscriminate antibiotic use is so dangerous: Emergence of resistance in commensal bacteria (see Box 6-3) can serve as a starting point for mobilization of resistance that eventually reaches pathogens.

Box 6-3: Antibiotic Resistance in Commensal E. coli

Analysis of *E. coli* taken from stool samples of preschool children in Bolivia and Peru illustrates the extent to which some environments are contaminated with antibiotics. (*E. coli* is a normal inhabitant of the digestive system.) An initial sampling in 2002 revealed a high prevalence of resistance to some agents (see Table 6-1). That level was still high 3 years later; moreover, fluoroquinolone (ciprofloxacin) resistance almost doubled between 2002 and 2005, and resistance to an advanced β-lactam (ceftriaxome) increased by 17-fold. Although the clinical significance of this resistance is not known, resistance genes can move from *E. coli* to other bacteria. Surveys indicate that the children did not receive direct treatment with ciprofloxacin or the β-lactam. The source of the resistant bacteria in this particular study is unknown, but an earlier study from Spain[146] attributed much of it to use of antibiotics with food animals. Part of the Spanish problem may have arisen from clandestine laboratories producing substandard drugs.

Table 6-1 Prevalence of Antibiotic Resistance in Preschool Children in Peru and Bolivia[148]

Antibiotic	Percent of Isolates That Were Resistant*	
	2002	2005
ampicillin	95	96
tetracycline	93	93
chloramphenicol	70	69
streptomcyin	82	92
kanamycin	28	29
amikacin	0.4	0.1
ceftriaxome (β-lactam)	0.1	1.7
ciprofloxacin (fluoroquinolone)	18	33

* The study involved a total of 3,000 children.

In still another study, volunteers were administered ciprofloxacin, and then stool samples were analyzed for quinolone-resistant *E. coli*. Resistant organisms were recovered from about one quarter of the subjects.[147] Those organisms may serve as a reservoir from which drug resistance passes to other bacteria.

Integrons Gather Genes into an Expression Site

Integrons are regions of DNA that gather together other DNA regions having the potential to encode proteins. Integrons then convert those regions into functional genes by placing them next to an active promoter, a region of DNA where RNA polymerase binds and begins making mRNA from an adjacent gene. Integrons do not move themselves, but instead they bring relatively small gene cassettes into an insertion site located next to a gene encoding a recombinase (integrase). To be moved into an integron, a region of DNA needs to have only a sequence that is related to the nucleotide sequence at the insertion site. Many of the "procured" genes encode proteins responsible for antibiotic resistance. Consequently, integrons, by accumulating sets of resistance genes, confer multidrug resistance (see Box 6-4). Integrons are involved in resistance to aminoglycosides, chloramphenicol, trimethoprim, rifampicin, erythromycin, fosfomycin, lincomycin, antiseptics of the quaternary ammonium family, and all known β-lactams.[149]

Integrons can move when they are located inside a transposon. Mobile integrons are found in many clinical isolates of multidrug-resistant bacteria and are a special problem with urinary infections. An example was reported from Uruguay in which 104 patient samples were examined.[150] Forty-six isolates were multidrug resistant, and 33 contained integrons (most of the integron-containing isolates were also multidrug resistant); one of the *Klebsiella pneumoniae* isolates contained 2 integrons and was resistant to 8 antibiotics. Nucleotide sequence analysis of some of the integrons revealed a complex history involving insertion into a transposon and homologous recombination between transposons. This type of study emphasizes how dynamic microbial DNA can be, moving pieces from one organism to another, inserting DNA pieces into other DNA molecules, and forming new combinations through genetic recombination.

Integrons are found in both Gram-positive and Gram-negative bacteria, suggesting that they may have an ancient origin. Nucleotide sequence analysis also reveals that they have combined genes from many different bacteria. One line of evidence derives from codon usage. During protein synthesis, amino acids are joined to tRNA molecules, one amino acid per tRNA. Most amino acids can bind to several tRNA types (one amino acid per tRNA molecule). Each type of tRNA has a different anticodon; consequently, each type recognizes a different codon in mRNA. For a particular amino acid, a given bacterial species tends to use one type of tRNA more often than the others. Therefore, mRNA shows a bias for certain codons, a bias characteristic of the bacterial species. That bias is called codon usage. Analysis of codon usage for genes assembled by integrons reveals a wide diversity, indicating diverse origins for present-day integrons.

Box 6-4: Integrons

Integrons contain three essential features that enable genes for resistance and other adaptive features to be gathered into a single region:

- A gene (*intI*) encoding an integrase enzyme that facilitates insertion of a gene into the integron
- A primary integration site (*attI*) where insertion occurs
- A strong promoter (P_c)

The integrase protein causes gene cassettes from DNA molecules to insert at *attI*, downstream from P_c such that promoter P_c drives expression of the captured gene. Multiple genes can be placed under control of P_c, creating a multidrug-resistance cassette. More than 70 different resistance cassettes have been identified that encode resistance to all β-lactams, all aminoglycosides, chloramphenicol, and many other antimicrobials. Each resistance cassette is bounded by short, direct nucleotide repeats that are targets for the recombination process. Integrons enable bacteria to stockpile resistance genes. (Some integrons, called superintegrons, have gathered as many as 200 genes.)[149] Analysis of a global collection of *Salmonella enterica*, a cause of food-borne disease, revealed an association of integrons with multidrug resistance.[151]

Genomic Islands Help Create Pathogens

The availability of nucleotide sequence information for many bacterial species has revealed that bacterial genomes contain large blocks of genes that differ from the surrounding DNA. For example, the ratio of AT to GC base pairs, which is characteristic of a chromosome, can be drastically different within a block. Genes within these regions can also show a codon usage preference that differs from the bulk of genes in the chromosome. These blocks are called genomic islands, or in some cases pathogenicity islands. A well-characterized example is found in *Enterococcus faecalis*.[152] This pathogenicity island, which is 150,000 base pairs long, contains 129 open reading frames (regions capable of serving as genes) that are likely to encode a toxin and other proteins that increase the capability of the bacterium to colonize the human gastrointestinal tract. The island also contains genes likely to be involved in conjugation, a variety of insertion sequence elements, and phage genes related to integrase and excisionase. Moreover, a portion of the island can transfer to other cells.[152] It is generally believed that some highly virulent bacterial strains arose when their genomes acquired mobile genetic elements carrying virulence genes.

Plasmid Enzymes Can Be Inhibited

During bacterial conjugation, an enzyme called relaxase introduces a single-strand break (nick) into the plasmid DNA at the spot (*oriT*) where transfer begins. The atomic structure of relaxase has been determined by X-ray crystallography, and the active part of the enzyme has been identified. It contains a pair of tyrosine amino acids that form covalent bonds with DNA during the DNA breakage and transfer process. Small molecules called phosphonates inhibit the relaxase, thereby blocking conjugation. Surprisingly, the phosphonates selectively kill plasmid-containing cells. Because relaxase is not required for cell survival, the phosphonate probably creates a toxic event rather than simply removing an essential cellular component.[153] One possibility is that a phosphonate-relaxase-DNA complex forms that triggers a cascade of lethal reactive oxygen species, as is seen with quinolones that trap gyrase on DNA.[154] Work on this type of process may eventually lead to the control of horizontal transfer.

Perspective

Our relationship with the living world is constantly changing. Human population growth and technology alter ecosystems, and other organisms of the Earth adapt as best they can. Some pathogens, such as those that cause smallpox and typhus, fade into the background, whereas others come forward. Examples of the latter are the human immunodeficiency virus and community-associated MRSA. Some changes we can easily understand: a vaccination program eradicated smallpox, and good hygiene plus delousing with DDT knocked typhus out of the headlines. Origins are sometimes unclear, as with CA-MRSA. As we expose bacteria in our bodies and in our surroundings to massive amounts of antibiotics, we strongly favor the growth of microbes with resistance genes. Those resistance genes move horizontally, that is, from one strain to another and from one species to another. Thus, antibiotic use favors microbes that effectively transfer resistance genes; strains that accumulate resistance features then spread within human populations. We can do little but watch the spread occur. In the next chapter, we consider how pathogens disseminate and spread disease.

Chapter 7

Transmission of Resistant Disease

Summary: Transmission of antibiotic resistance occurs through the spread of resistant pathogens from one person to another. Disseminated resistance is easily seen as disease outbreaks that are unresponsive to particular antibiotics. Resistant pathogens are typically as effective at causing disease as susceptible ones; consequently, transmission of resistance is similar to transmission of disease. Controlling outbreaks of infection relies on local surveillance and interventions such as patient isolation, identification of effective antibiotics, and in some cases, vector control. For most diseases, hand hygiene is crucial. Some pathogens are spread by air (tubercle bacillus, influenza virus); some by contact (*S. aureus*); some by food and water (Salmonella, *Vibrio cholerae*); some by exchange of body fluids (some hepatitis viruses); and some by insects (malaria parasite, yellow fever virus). Tuberculosis is an example of a treatable airborne infection that has been exhibiting outbreaks of resistant disease around the globe. Malaria is an insect-borne example that is endemic in many tropical countries. With industrialized countries, the widespread multidrug-resistant diseases of the near future are likely to include MRSA infection, gonorrhea, and hospital-associated diarrhea.

We treat with antibiotics, resistant pathogens emerge, and those pathogens then spread from person to person. That spread causes widespread loss of antibiotic effectiveness. We begin this chapter with a brief discussion of disease transmission and infection control. Then we discuss a variety of diseases to illustrate how crisis situations are addressed.

Spread of Pathogens Is Highly Evolved

As long as a resistant infection is limited to its initial patient, disease can be controlled by isolating the patient. But when isolation is not achieved and the pathogen spreads, resistance also spreads. Resulting infections generally fail to respond to the antibiotic, and an outbreak of drug-resistant disease can occur. Indeed, outbreaks are what usually come to mind when we hear the term antibiotic resistance in reference to situations in hospitals.

Transmission of resistant pathogens follows the same paths as transmission of susceptible pathogens: through the air, in food and water, by direct contact with infected persons, and via insect vectors. Consequently, controlling resistant outbreaks employs the same infection-control practices normally associated with disease. But the sense of urgency is much greater, because the usual antibiotic-based cures won't work—we must switch to other agents, if they are available.

Infection Control as Local Crisis Management

Hospitals and health departments have staff responsible for identifying infection outbreaks, defining the pathogen involved, and developing solutions. Monitoring reportable diseases, such as tuberculosis, is mandatory. However, keeping track of many other diseases is discretionary; consequently, institutions can vary considerably in the quality of their infection control programs. One of the most visible activities is controlling outbreaks of resistant disease. Usually this means containment: patient isolation, special glove and gown procedures for anyone entering a particular patient's room, and sometimes use of rooms with controlled, filtered airflow. If possible, the disease is suppressed by identifying antibiotics that are effective. Often, outbreak control is performed when many lives and considerable money is at stake. Infection control programs also monitor and control antibiotic use in hospitals and long-term care facilities to preempt resistance. These institutions, including jails and prisons, can be breeding grounds for resistant diseases that then escape into the community.[155] Some of the specific tasks of hospital control programs are sketched in Box 7-1.

Box 7-1: Hospital Antibiotic Control Programs

Hospital antibiotic management teams are responsible for a variety of activities.[156] First, formularies are set up to exercise broad control over prescribing by limiting the antibiotic classes and members of classes approved for hospital use. These decisions are based on cost, efficacy, adverse effects, and resistance. Second, some antibiotics are made available only for specialty uses, and prior approval may be required for restricted-access antibiotics. Third, management teams institute standard treatment paths (clinical pathways) for certain infections to reduce decision making and presumably the chance for error. These pathways include stop orders on prophylaxis cases. Local surveillance information and expert clinician input guides treatment. Fourth, the teams review patient outcomes on a case-by-case basis and keep tabs on institutional overuse. Fifth is antimicrobial streamlining, a process in which antimicrobial therapy is matched to susceptibility data and narrow-spectrum agents are encouraged when appropriate. A sixth activity involves review of conversion from intravenous treatment to oral antibiotics by identifying appropriate doses and agents. This reduces the number and duration of skin breaks. A seventh activity concerns provider education. That includes mailings, direct education, and peer review of prescribing and hand hygiene practices.

In the community, health departments watch for disease outbreaks. Sometimes they limit the spread of disease by tracking persons who have been in contact with an ill person. Such is the case with tuberculosis. At other times, health department personnel follow tainted food to the source. Surveillance reveals the problems, and containment teams develop solutions.

In the remainder of the chapter, we describe examples of pathogen transmission and the response of infection control specialists. A broader view of the effort to control antibiotic resistance is presented in the next chapter. There we discuss surveillance that may span years and methods for determining resistance. The broader view reveals patterns of increasing prevalence of resistance, erosion of antibiotic effectiveness, and the need for policy changes and new antibiotics.

Tuberculosis Is Airborne

We begin our examples of disease transmission with tuberculosis. In industrialized nations, tuberculosis serves as an example of successful infection control; in some developing countries it illustrates the consequences of failure.

When *Mycobacterium tuberculosis* (see Figure 7-1) gets deep inside a person's lungs, human scavenger cells called macrophages engulf it. These amoeba-like cells normally protect us from disease by killing pathogens, but *M. tuberculosis* survives. For reasons that are poorly understood, about 90% of infected persons, if otherwise healthy, either clear infection or drive *M. tuberculosis* into a dormant state. In the dormant state, *M. tuberculosis* can remain hidden for decades, ready to resume reproduction when the infected person's immune system is impaired. If the bacteria fail to shift into the dormant state or when they come out of it, they reproduce. Eventually, they create cavities in the patients' lungs. Patients with active tuberculosis have a persistent cough; sometimes they cough blood and lose weight as they are consumed by the bacteria. (For centuries, tuberculosis has been called consumption.) Death frequently follows if antibiotic treatment is unavailable. (The death rate is about 50% for immune-competent patients.)

Tuberculosis is spread by air (see Box 7-2). Coughing by a tuberculosis patient produces tiny droplets that contain infectious bacteria. These droplets quickly dry and form droplet nuclei that can remain airborne for several hours. If another person inhales the droplet nuclei, that second person can become infected. *M. tuberculosis* now infects about one-third of the world's population.

Figure 7-1 *Mycobacterium tuberculosis*. Scanning electron micrograph of *M. tuberculosis* at a magnification of 15,549 times.

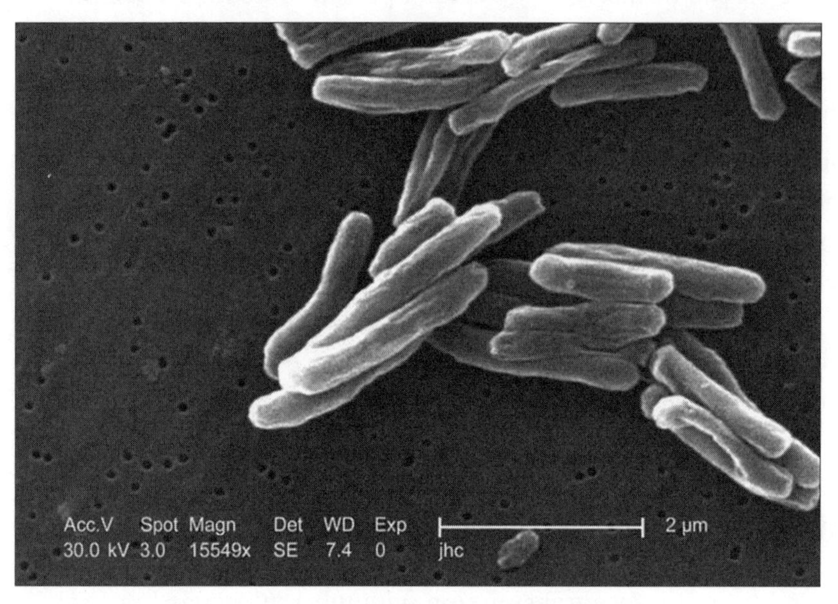

Source: Public Health Image Library #8438; photo credit, Janice Haney Carr.

Box 7-2: Guinea Pig Test for Tuberculosis

The airborne nature of tuberculosis was established in the 1950s by exposing a colony of guinea pigs to air from the rooms of tuberculosis patients and finding that the animals got tuberculosis. In a recent experiment, air from a ward of HIV-positive tuberculosis patients was passed to guinea pigs housed on the roof of the hospital. Analysis of *M. tuberculosis* from the patients and the infected animals revealed that a small number of patients were highly contagious, whereas other patients appeared less contagious. In this study, the infectious patients had been inadequately treated. Because antibiotic-resistant forms of tuberculosis are difficult to treat, proper ventilation or isolation of hospital wards and *waiting rooms* for tuberculosis patients is essential.[157]

Industrialized countries control tuberculosis by mandating case reporting, by directly observed therapy with multiple antibiotics, by patient isolation, and by special air handling systems for rooms where tuberculosis patients are housed. Laws are in place to quarantine patients if necessary. In some cases, hospital staff wear personal respirators or masks. The most notorious outbreaks of tuberculosis have occurred in hospitals housing large numbers of sick persons with weak immune systems. (Persons who are infected with HIV and untreated for the virus progress rapidly to active tuberculosis when infected with *M. tuberculosis.*)

We have several ways to detect tuberculosis in its early stages. One involves a simple skin test that reveals an immune response to infection (see Box 7-3). Another is chest X-ray. During disease, an infected person's body responds by walling off infected lung tissue, forming spots called granulomas. When large enough, granulomas can be seen using X-rays. Late stages of disease are recognized by fever, wasting, and coughing blood.

Box 7-3: Testing for Exposure to M. tuberculosis

When a small amount of *M. tuberculosis* extract (a preparation of broken cells) is injected under a person's skin, it produces a raised region after a day or so if the person has been infected with *M. tuberculosis*. If the person has not been infected, the skin remains flat at the injection site. In this way, the tuberculin skin test serves as a warning of infection. A vaccine for tuberculosis, prepared using the attenuated strain *M. bovis* BCG, is often administered in countries where the incidence of tuberculosis is high. Vaccination protects young children (under 5 years old) from milliary tuberculosis, a condition in which *M. tuberculosis* invades beyond the lungs. The protective effect of vaccination is marginal with older children and adults. One of the consequences of vaccination is production of a cellular immune response that reacts with *M. tuberculosis* extract, even years after vaccination. For persons who have been vaccinated, a positive result to the skin test may indicate vaccination rather than infection by *M. tuberculosis*. Thus, vaccination eliminates use of the skin test for detecting tuberculosis. For that reason some countries having low tuberculosis incidence, such as the United States, do not vaccinate.

Two new tuberculosis tests recently became available for persons who already test positive by the skin test. These tests, called interferon-γ release assays, measure the production of interferon-γ by lymphocytes exposed *M. tuberculosis* proteins that are absent from *M. bovis* BCG. Consequently, persons who were vaccinated but free from *M. tuberculosis* infection will score negative. This test, like the skin test, cannot distinguish among latent infection, active disease, and treated tuberculosis.

The tuberculosis tests enable us to stay ahead of the pathogen by providing antibiotic prophylaxis to persons testing positive. Decisions about tuberculosis prophylaxis rely on surveillance data and epidemiological information, because immunological tests do not reveal information about drug resistance. (Resistance can be determined only after latent infection converts to active disease, because only then can the pathogen be obtained for testing.) If infection with an antibiotic-resistant strain of *M. tuberculosis* is probable because the source of *M. tuberculosis* is a patient with a drug-resistant infection or because local strains are drug resistant, prophylaxis with the usual agents, isoniazid or rifampicin, may be ineffective. In such cases, another drug must be chosen.

Health departments in the United States track tuberculosis cases. However, they do not follow infection in the absence of active disease (positive skin test with negative chest X-ray), because persons with latent tuberculosis are not infectious.[158] Nevertheless, such persons may serve as an early warning sign that someone else in the community is spreading *M. tuberculosis*. Moreover, the skin test enables health departments to evaluate their tuberculosis control programs. For example, homeless shelters are thought to be a reservoir for tuberculosis and therefore good places to monitor control efforts. In a New York City study, the frequency of skin-test-positive homeless persons dropped from 30–40% in 1992 to about 10% in 2006.[158] This good news is tempered by anecdotal data recently revealing that some middle-class American teenagers are unexpectedly exhibiting positive skin tests. The magnitude and source of this problem is currently unknown.

Although antibiotics cure tuberculosis, cure is not easy to achieve. Even during active disease, part of the pathogen population is probably dormant and only marginally susceptible to antibiotics. The current practice is to treat for 2 months with four antibiotics (isoniazid, rifampicin, ethambutol, and pyrazinamide) and then 4 months with two drugs (isoniazid and rifampicin). Some of the drugs have serious side effects, which makes adhering to the treatment regimen difficult. Moreover, drug action causes patients to feel better; consequently, they tend to stop taking the antibiotics. If treatment stops too soon, residual bacteria reproduce. Patients feel ill again, and drug treatment is resumed. Stopping treatment permits bacterial outgrowth and the opportunity for new spontaneous mutants to arise; resumption of treatment places selective pressure on the bacterial population. Indeed, physicians generally attribute the emergence of resistance to patients who fail to adhere closely to the prescribed regimen. The adherence problem is reduced by directly observed therapy (DOT), a procedure in which healthcare workers are sent daily to patients' homes to assure that the medications are taken. DOT is now a worldwide practice for tuberculosis treatment.

Prior to treatment, a tuberculosis infection can contain more than a billion bacterial cells. For each of the antibiotics used with tuberculosis, the large bacterial population can include resistant mutants. For some drugs, a thousand mutants may be present. When only one of the prescribed drugs is present and active, which can happen for a variety of reasons, the small mutant subpopulation will amplify. Those mutants grow in the presence of the drug, even though the bulk of the population may be killed. Eventually the mutants dominate the population. When the bacterial population resists one drug, the chance increases that exposure to a second agent will occur under conditions in which it is the only active compound. Over time and across millions of patients, resistance mutations gradually accumulate in the bacterial populations. (Some of the *M. tuberculosis* isolates in our collection are resistant to eight different antibiotics.[159])

When a strain of *M. tuberculosis* acquires resistance to the two main antituberculosis drugs, currently rifampicin and isoniazid, the strain is called multi-drug resistant (MDR), and the disease is called MDR tuberculosis. MDR tuberculosis is difficult to cure, because the remaining agents are not as effective as rifampicin and isoniazid. Accumulation of additional resistance mutations leads to extensively resistant (XDR) tuberculosis (MDR tuberculosis that is also resistant to a fluoroquinolone and an injectable drug, such as kanamcin, amikacin, or capreomycin). Several other agents are available for treatment of XDR tuberculosis, but they are not very effective. Moreover, their side effects can be severe. With some patients, we reach the end of the line: completely drug-resistant tuberculosis (CDR-TB). At each step, the disease can spread to other persons through the air.

In the 1970s and early 1980s, the number of tuberculosis cases declined. The medical community and funding agencies lowered their guard with bacterial diseases in general and with tuberculosis in particular. But in 1981, the first cases of AIDS were recognized, and within a few years HIV-positive persons were common in institutional settings (hospitals, shelters, and prisons). Because HIV infection permits rapid growth of *M. tuberculosis*, tuberculosis became a nosocomial (hospital) infection. In the late 1980s, a strain of *M. tuberculosis* that carried resistance to the four major antituberculosis agents entered the immune-compromised population of New York City and spread. Patients began to die, sometimes within only a few months after being infected.

New York City prisons and hospitals built special rooms in which air flow is controlled to prevent spread of the pathogen. Negative pressure is maintained inside the containment room so that air flows in from the hallways; room air is then exhausted through filters designed to capture the infectious bacteria. The

MDR tuberculosis outbreak was eventually suppressed by air control, isolation of patients, and directly observed therapy using second-line antituberculosis agents. (Thirty-eight hundred cases of tuberculosis were reported in New York City in 1992; of these about 30% were MDR-TB.[160]) MDR-tuberculosis also appeared in other countries, some of which made cost estimates. For the British healthcare system, the cost for one MDR tuberculosis case was 60,000 pounds sterling (1996–99), which was about 10 times the cost of a susceptible case.[161]

The 1990s also saw the former Soviet Union experience a severe tuberculosis problem in its prisons. (Crowded conditions enabled the disease to spread among prisoners.) Poor diet and inadequate, sporadic antibiotic treatment spurred the development of MDR tuberculosis. Prisoners then returned to the general population after serving their sentences. Follow-up tuberculosis treatment was grossly inadequate, and MDR tuberculosis began to move through Russia and adjacent countries. Siberia became a notorious tuberculosis locale; one of its cities, Tomsk, now serves as a training center for tuberculosis healthcare workers (see Box 7-4).

In the late 1990s, HIV started its spread through Eastern Europe, exacerbating the MDR tuberculosis problem. (By 2002, the prevalence of MDR tuberculosis was 23% in Lithuania compared to 2–3% worldwide.[162]) As expected, XDR tuberculosis emerged, and in some localities, such as the Orel region of Russia, it represents 20% of the cases.[21]

Box 7-4: Training TB Workers in Russia

The Russian prison system has been a global hotspot for MDR tuberculosis. In the mid-1990s, a region in Siberia (Tomsk Oblast) became the focus of international efforts to combat MDR tuberculosis. MERLIN (Medical Emergency Relief International), Partners In Health, and the Public Health Research Institute, funded largely by George Soros, initiated an effort to reduce disease incidence by supplying antituberculosis agents and technical support.[155] Tomsk is now an international training center for healthcare workers trying to combat MDR tuberculosis. In Tomsk, the focus is on home and outpatient treatment provided by nurses and community workers. (Placing tuberculosis patients in hospitals is thought to increase the spread of disease.) All patients are registered in a database that enables medical personnel to track them and ensure that they stay the course. So far, HIV has not been a major factor in Tomsk, which makes it different from South Africa, another tuberculosis hotspot. Getting high-quality antibiotics to patients on a regular basis is a common challenge in regions where tuberculosis prevalence is high. (Such regions are identified by the World Heath Organization; in 2007, the tuberculosis burden [new cases per year] was estimated to be in South Africa 948; China 98; Russia 110; India 168; and United States 4.2 per 100,000 population.)

HIV disease also spread throughout sub-Saharan Africa. Tuberculosis was already widespread, and dissemination of HIV reactivated latent tuberculosis. Where antituberculosis drugs were applied, MDR tuberculosis emerged and spread. By the early 2000s XDR-tuberculosis began to make headlines. In one rural center (Kwa Zulu Natal, South Africa), HIV-infected persons visited a local clinic for treatment and caught XDR tuberculosis from other patients. Within slightly more than 2 weeks of diagnosis, 49 out of 52 infected persons died.[20]

The United States and many other industrialized countries have kept tuberculosis under control with aggressive drug treatment. In the United States, roughly 14,000 cases of tuberculosis are reported per year, half of which involve foreign-born persons. Fewer than one percent are XDR. These low numbers mean sparse media attention. However, we cannot let down our guard. Global travel, particularly on airplanes where everyone breathes the same recycled air, puts each of us at risk. We also need to pay attention to pockets of tuberculosis, such as homeless shelters (see Box 7-5).

Box 7-5: Tuberculosis and the Homeless

On a worldwide basis, about 100 million people are considered homeless. A report in the year 2000 indicated that tuberculosis is much more prevalent among the homeless. For example, in San Francisco, 270 tuberculosis cases per 100,000 inhabitants were found among the homeless compared to only 4.6 per 100,000 in the total U.S. population. Outbreaks tend to occur in large clusters. (Homeless shelters are associated with disease transmission.) Because the homeless population is quite mobile, identifying infected persons and delivering months of treatment is often difficult.[163] Nevertheless, an aggressive public health effort in New York City has been effective: In 1992, the calculated frequency of tuberculosis among the homeless was 1,500 per 100,000 total population which dropped to 170 per

continues

100,000 in 2004. (In 2007, the value for the general New York City population was 13 per 100,000 and for the United States as a whole it was 4.4 per 100,000.[158])

Airborne Viruses

Many respiratory viruses move from person to person through small droplets created by coughing, sneezing, and talking.[164] We can also pick up viruses on our hands. When we touch our faces, the viruses then pass to nose and mouth. Shaking hands passes the virus. Many of the precautions mentioned for stopping *M. tuberculosis* transmission are useful with viruses: masks, gowns, gloves, and controlled airflow. Among the more notorious airborne viruses are the cold viruses, influenza virus, and severe acute respiratory syndrome (SARS). Influenza and SARS are major killers that merit additional discussion. Antibiotics are available for influenza, which we discuss in Chapter 11, "Influenza and Antibiotic Resistance." SARS is inherently drug-resistant, that is, we have no antibiotic for it. This disease broke out in China in 2002, and the virus quickly spread to Hong Kong. From there, it moved to Toronto where hospital workers carried the disease to the community (see Box 7-6). Stringent infection control procedures eventually brought SARS under control.

Box 7-6: Severe Acute Respiratory Syndrome—No Known Drugs

In his 1969 book, *The Andromeda Strain,* Michael Crichton described an alien life form brought back to earth from a secret government satellite. Deadly disease spread through a small Arizona town, wiping out much of the population in the blink of an eye. The virus-like disease was like nothing previously seen on earth. Science fiction has a perverse way of becoming reality. From November 2002 through July 2003, much of the world lived out this techno-thriller, as an unknown virus causing deadly atypical pneumonia spread rapidly from country to country. When the outbreak subsided, 8,096 known cases of infection and 774 deaths had occurred (case-fatality rate of almost 10%). The virus causing Severe Acute Respiratory Syndrome (SARS) had disseminated from China's Guangdong province to 37 countries around the world. The virus, a type of coronavirus, spread rapidly like pandemic influenza, moving easily from person to person. In China, streets were filled with countless thousands donning masks for protection. The airports swarmed with "temperature monitors" who searched for feverish disease carriers.

Unfortunately for the city of Toronto, precautions came too late. Shortly after the index case died, a sick family member spread the virus to a key hospital. From there, the virus spread rapidly, especially among healthcare workers, and the World Health Organization warned travelers to avoid Toronto. Existing antiviral drugs were ineffective. The ultimate resistant "bug" spread rapidly and left social and economic devastation in its wake. The only meaningful approach was to break the chain of transmission by effective isolation and quarantine of patients, a remedy that has rescued civilization from plagues for centuries. In a short time, the epidemic waned, and then it was over.

Digestive-Tract Pathogens

Diseases of the digestive tract, such as cholera and typhoid fever, usually spread by food and water contaminated by infected persons. Developed countries keep these diseases in check largely through sanitation systems, with antibiotics being used to clear up sporadic cases. (Raw shellfish are a common source of sporadic cases of cholera.[165]) Outbreaks of cholera and typhoid are often associated with man-made or natural disasters, usually in underdeveloped countries. Industrialized countries worry more about hospital-borne pathogens. Species of *Enterococcus* are common inhabitants of the human digestive tract that are responsible for a variety of infections. Because *Enterococcus* is often antibiotic resistant, considerable attention is given to halting its spread through hospitals and long-term care facilities (see Box 7-7).

Box 7-7: Hospital Outbreak of Vancomycin-Resistant Enterococcus

Vancomycin-resistant *Entercococcus faecium* is easily transmitted among patients. Hospitals control transmission by isolating patients and by instituting rigorous hygiene-related infection-control practices. In late 2004 and early 2005 an outbreak occurred in a German hospital.[166] The incident ultimately involved more than 100 patients who were kept in isolation for more than 2,600 patient-days. The estimated cost for suppressing the outbreak exceeded 1 million euros. Had this outbreak been addressed sooner (4 months were required for microbiologists to convince administrators that the patients represented an outbreak rather than the sporadic cases that occur routinely), cost and suffering could have been reduced. One challenge is to reliably detect an outbreak early.

Hand washing has been the major response to intestinal infections, and food handlers have long been the focus of hand-washing programs. More recently, doctors and nurses have been implicated in transmitting a variety of microbes from one patient to another. (Keyboards of computers are an obvious pathogen reservoir.) Even mental hospitals have "Clean hands save lives" signs posted throughout. Although washing procedures may seem simple, multiple hand washings can cause skin irritation. Moreover, they require time from a busy schedule. Plastic gloves provide personal protection from diseases spread by hand contact, but they also spread pathogens if not properly discarded immediately after each use. Even medical instruments, such as stethoscopes, are found to be contaminated with pathogens and require frequent cleaning.[167]

Direct-Contact Pathogens

Some pathogens are spread by skin-to-skin contact. In Chapter 1, "Introduction to the Resistance Problem," we briefly described situations involving *S. aureus*. Although hospital outbreaks of MRSA have been common since the 1970s, community-associated forms began to emerge only in the late 1990s. Examples involving military recruits and sports teams are consistent with transmission occurring by skin-to-skin contact and sharing personal items (see Box 7-8).

Box 7-8: Spread of MRSA

Military recruits at training facilities are at risk for community-associated (CA) MRSA. Between October 2000 and June 2002, a large U.S. military facility recorded 235 cases of CA-MRSA. In November 2002, military authorities implemented a variety of hygienic measures including an emphasis on hand washing and showering. In addition, sharing personal items was prohibited, and antibiotic therapy was instituted to eliminate nasal colonization. The outbreak ended in December 2002. In many of these cases, infections were on arms and legs where skin abrasions were expected from training exercises.[168]

Another example surfaced in September 2005 when five members of the St. Louis Rams professional football team reported MRSA infections at turf-abrasion sites. The abscesses were large (more than 2 inches [5 cm] in diameter) and required surgical incision and drainage. Molecular analysis of the infecting bacteria showed that all cases were due to USA300, which was common in the community. A variety of infection control procedures were instituted. For example, hand washing was encouraged using bactericidal agents, as was showering before whirlpool treatments. The common practice of towel sharing was stopped, and weight-training equipment was regularly sanitized. In addition, antibiotic treatment was used on infected players. Although MRSA appeared to spread to opposing teams during games, analysis could not distinguish between the infecting strains and strains present elsewhere in the community.[169]

A third example was observed with a Dutch soccer team. In June 2005, several players noticed soft-tissue infections, and in October, one member of the team was hospitalized for an MRSA infection. Screening of team members and their close associates (56 persons) revealed MRSA in nine players and two roommates. DNA fingerprinting of the MRSA isolates showed that they were identical, consistent with person-to-person transmission. Members of the team were advised to share no personal item, to use disposable towels after showering, and to place a disposable towel on locker room benches before sitting. Ventilation was improved in the team locker room, which was also cleaned more frequently. These strategies, plus antibiotic treatment, stopped the outbreak.[170]

Because nasal surveillance appears to be effective with hospital-associated *S. aureus*, a similar strategy was examined with a professional football team.[171] In this example, none of the players exhibited nasal colonization at the beginning of the season. Nevertheless, five cases of infection occurred during the season, and at the time of infection none of the five exhibited nasal colonization. Thus, screening for nasal colonization may not be an effective strategy for predicting disease.

Arthropod-Borne Pathogens

Many serious pathogens are spread by insects and ticks that bite. Among the more notable are malaria, sleeping sickness, yellow fever, and West Nile Fever. Malaria is the most affected by drug resistance. This disease is endemic to more than 100 countries, with transmission occurring throughout the year in the tropics and during summer months in temperate climates. *Plasmodium falciparum*, a major cause of malaria, exhibits widespread resistance to chloroquine. By 2008, the drug was effective only in Mexico, the Caribbean, parts of Central America, East Asia, and some Middle Eastern countries.[172] On a worldwide basis, drug-resistant malaria is becoming such a large problem that health officials no longer describe outbreaks. Instead they speak of areas of endemic resistance. General features of drug-resistant malaria are sketched in Box 7-9.

Box 7-9: Antibiotic-Resistant Malaria

Human malaria is caused by four species of *Plasmodium*. *P. falciparum* is the most prevalent in Africa south of the Sahara. *P. vivax* causes 40% of the cases worldwide; it is the dominant cause outside Africa. *P. malariae* and *P. ovale* are less prevalent, but they have worldwide distribution and are especially prevalent in tropical areas of Africa. Occasionally humans are infected with monkey pathogens, such as *P. knowlesi*. *P. falciparum* causes the most serious disease; consequently, it merits the most attention. Identification of the causative pathogen can be important, because the species vary in susceptibility to available drugs. Chloroquine resistance in *P. falciparum* was first noticed in Thailand and Colombia in the late 1950s, a dozen years after the drug was introduced. In 1978, chloroquine resistance emerged in East Africa and spread westward throughout the 1980s. By 1980, all endemic areas of South America were affected, and by 1989 most of Asia and Oceania reported resistant parasites.

Several practices are thought to contribute to the emergence of resistance. One is the use of treatments that fail to completely eliminate the parasite. Assessment of therapy, which was commonly performed after 14 days of treatment, was mainly for loss of symptoms. Surviving parasites may not regrow to high enough numbers to cause symptoms by 14 days; thus, they may be overlooked. Effects of regrowing parasites may also be mistaken for re-infection. (Three infectious mosquito bites per day is common in areas where transmission is high, which makes re-infection common.) Consequently, an inadequate treatment regimen was deemed adequate. These practices led to cycles of drug treatment, which we know from bacterial studies selectively enrich mutant pathogen subpopulations.

Other factors contributing to emergence of resistance are substandard drugs, poor adherence to therapy, and uneven drug availability. Moreover, in regions where transmission is high and infection is frequent, many adults are semi-tolerant: Consequently, they would typically receive partial treatment regimens to remove symptoms. In 2006, the World Health Organization reset assessment time to 28 days, encouraged parasitological assessment, discouraged monotherapy (use of only one antibiotic), and suggested changing agents when resistance prevalence exceeds 10%.

Malaria affects young children most severely. Persons who survive to adulthood have partial immunity that enables them to live with the parasite. However, that tolerance disappears quickly when those persons move to countries where repeated exposure to the parasites does not occur. Consequently, travelers from industrialized countries, even those with previous exposure to malaria, lack immunity; they tend to suffer much more serious disease than long-term residents of areas where malaria is common. Strategies for short-term travelers are discussed in Box 12-6.

Another arthropod that spreads disease is the human body louse. This insect transmits typhus, which is caused by an obligate bacterial pathogen of the group called rickettsia.[173] The rickettsial species that causes typhus invades epithelial cells of the louse digestive tract, reaching human skin via louse feces. Infection of humans then occurs through openings in the skin caused by louse bites. In humans, typhus causes severe headaches, high fever, rashes, and often death. (60% fatality if untreated). The lice, which are obligate parasites of man, are rarely found on persons in industrialized countries. (The last typhus outbreak in the United States occurred in 1922, before antibiotics were available.) However, crowded living conditions associated with war and famine still lead to outbreaks of typhus in Africa. (In 1997, an outbreak involving 100,000 persons occurred during a civil war in Burundi.) Rickettsia are readily controlled by tetracyclines. Although antibiotic resistance is unlikely to arise from the sporadic cases currently occurring, drug-resistant forms could emerge from inadequate treatment. Persons who survive typhus can carry a latent form of the pathogen that reactivates and causes disease during times of stress or immune compromise. Thus, resistant rickettsia could persist among infected humans. Because dried louse feces containing rickettsia can lead to airborne infection, resistant forms could conceivably spread. Fortunately, insecticides have been successful with louse- and flea-borne diseases. (World War II era refugees were commonly deloused by DDT dust.) Body lice die in 5 days if they fail to get a blood meal; therefore, louse-infested clothing can be set aside for a week to help control these organisms.

A variety of serious diseases are spread by ticks, which are arthropods closely related to insects. One of the more troublesome is Lyme disease. This infection is caused by a bacterium (*Borrelia*) that is currently susceptible to many antibiotics. Although resistance may conceivably emerge in individual patients, the resistant bacteria are not likely to spread to another person because the vector (deer tick) does not commonly travel from one person to another. To generate a resistance problem, we would need to treat deer or mice with antibiotics.

The obvious solution for the control of arthropod-borne disease is to suppress vector populations. Often, these are mosquitoes. Such an approach has been successful in industrialized countries: Both yellow fever and malaria are rare in the southern United States, a region where they were once common. Yellow fever is further controlled by a vaccine. In temperate, industrialized countries, control of surface water can be important, as indicated by an outbreak of West Nile Virus in California (see Box 7-10).

Box 7-10: Swimming Pools as Mosquito Breeding Grounds

West Nile Virus, which is spread by several species of the *Culex* mosquito, started in the Northeastern United States in 1999 and arrived in California in 2003. In 2007, Bakersfield health authorities reported an outbreak of 140 human cases, which was a 200% to 280% increase over recent years. (This unanticipated outbreak of an encephalitis-type virus was the largest in the area since 1952.[174]) The winter and spring months had been dry, and the population of the rural species of *Culex* was down. However, numbers of *C. pipiens*, an urban mosquito, had increased. Aerial searches for water sources revealed a large number of neglected swimming pools and ornamental ponds. Closer inspection showed that many were infested with mosquito larvae. These pools of water were blamed on the depression in the housing market and adjustable-rate mortgages that combined to dramatically increase the number of home foreclosures and abandonments. Early in the spring of 2008, health authorities treated swimming pools and sprayed insecticide on areas where adult mosquitoes were detected. These measures, plus continuing drought conditions, greatly reduced West Nile Virus transmission to humans.[175,176]

Blood-Borne Infections

Many viruses, including HIV-1 and several hepatitis viruses, are transferred by needle sharing and blood transfusions. Diseases associated with these viruses are a particular problem with intravenous drug users. They are also problematic in countries where the same hypodermic syringe is used with multiple patients. In the 1980s, before blood supplies were rigorously protected, many transfusion recipients acquired HIV through blood donated by infected persons. Moreover, healthcare workers occasionally caught the virus through accidental needle sticks. As antibiotic-resistant HIV becomes more common, it too will move among intravenous drug users through needle sharing. Our response to blood-borne pathogens has been to break the lines of transmission: Screen blood supplies, provide drug addicts with clean needles, and enforce safe use of needles by researchers and hospital staff. (Syringes are available that have a protective plastic sleeve that pulls over the needle after use.)

Multiple-Mode Transmission

Some pathogens seem to move around a hospital in a variety of ways. *Klebsiella pneumoniae* (see Box 7-11) provides an example that was reported in the popular press.[177] In 2000, the intensive care units of a major New York City hospital experienced a *K. pneumoniae* outbreak. Aggressive hygienic measures were instituted: rigorous hand washing, use of gowns and gloves, and application of disinfectants to all surfaces several times a day. However, infections continued to pop up until it was noticed that disposal of urine containers allowed splashing. That led to contamination of gloves and spread of the pathogen. Patients were moved out, and the rooms were disinfected. Patients were also isolated; visitors were forced to wear gloves and gowns. The outbreak subsided after 3 years. In the first year, half of the 34 infected patients died from infection, a number that emphasizes the seriousness of *Klebsiella.*

Box 7-11: Klebsiella pneumoniae

K. pneumoniae is a Gram-negative bacterium related to *E. coli* that causes pneumonia. Isolates resistant to all common antibiotics are an emerging threat to hospitalized patients. In many U.S. hospitals, *K. pneumoniae* harbors plasmids carrying genes for extended-spectrum β-lactamases (ESBLs). These enzymes confer resistance to penicillins and cephalosporins, but until recently they had little effect on carbapenems. An isolate resistant to carbapenems was recovered in North Carolina in 1996, and by the early 2000s, it had become a serious problem in New York City hospitals. In 2004, almost 25% of *K. pneumoniae* isolates were carbapenem resistant; by 2006 the number had risen to 36%. By 2004, the organism had been recovered in 20 states, France, Scotland, Israel, and China. This pathogen commonly colonizes digestive tracts of healthy persons, which allows it to spread easily. Indeed, patients can acquire drug-resistant *K. pneumonia* in a hospital and then carry it to the community in their digestive systems. We note that the ESBL-carrying plasmid has also spread to *E. coli*, *Salmonella enterica*, and *Pseudomonas aeruginosa*,[178] which makes these organisms more difficult to control.

Current hospital guidelines call for microbiology laboratories to test isolates for carbapenemase, a bacterial enzyme that contributes to resistance. When a carbapenmase-positive isolate is found, notification is sent immediately to infection control staff. Contact precautions are then implemented for the patient harboring the isolate. If the patient did not bring the pathogen into the hospital, an active search for the source is carried out. This can involve examining rectal swabs from all patients who had contact with the positive case or shared the same healthcare provider. When other positive cases are found, contact precautions are applied to them, and the search continues.

The *Klebsiella* problem, plus several of the examples previously presented in the chapter, illustrate the task of infection control experts: Identify outbreaks early, discover the cause, and institute corrective action. If possible, preempt outbreaks by assuring compliance with guidelines established by experts. For example, patients infected with pathogens that are easily spread by contact are placed in rooms where special precautions are expected of staff and visitors. Unfortunately, adherence to the guidelines is far from perfect (see Box 7-12).

Box 7-12: Controlling Resistant Infections Spread by Contact in Hospitals

One strategy to control pathogens spread by contact is to 1) identify the patients carrying contact-transmissible disease, 2) place signs on the rooms requiring contract precautions, and 3) supply rooms with gowns and gloves. Staff and visitors to the room are expected to don gowns and gloves, both of which are to be discarded upon exit. Hand hygiene is also expected at exit.

In a 2008 study, three areas of a major New York City hospital system were monitored for compliance to the precautions for patients with vancomycin-resistant enterococci, MRSA, *K. pneumoniae, C. difficile,* and *P. aeruginosa.*[179] In the study, more than 1,000 persons entered the posted rooms. One problem was that only 85% of the rooms qualifying for posting were posted. Another was poor adherence to the guidelines: only 20% for hand hygiene, 67% for gloves, 68% for gowns upon entry, 48% for hand hygiene upon exit, 63% for disposal of gloves, and 77% for disposal of gowns. More than 150 instances were observed in which staff/visitors potentially contaminated the hospital environment, mainly by failing to remove gloves upon exit from the rooms.

The term "hand hygiene" does not necessarily mean hand washing.[180] Scrubbing hands extensively can damage the skin and cause it to actually harbor more bacteria.[180] Alcohol-based cleaners have the advantage that they are rapidly acting, but they do not kill spores.

Perspective

Outbreaks of resistant pathogens are the most visible consequences of antibiotic practices that fail to keep mutant pathogen subpopulations small. Because resistant pathogens spread by the same routes as susceptible ones, we already know what to do: interrupt transmission. But often that must be done without our first-line antibiotics, which makes the task more difficult and expensive. Over the years we have made progress in the community through sanitation, vaccines, and vector control. However, that progress is countered by our increasing population density. Moreover, our populations now contain large numbers of persons whose bodies do not fight infections effectively. Among these are the elderly, transplant patients, and persons with immune deficiency diseases (AIDS).

With hospitals and long-term care facilities, we foresee considerable work for epidemiologists who specialize in infection control. As pointed out, their task is formidable, even for a behavior as basic as hand hygiene. As graduate students we were told about a mythical former member of the lab who wore gloves while working with radioactive isotopes but then neglected to promptly discard them when he was done. The supervisor would relate how other lab members walked around the lab with a Geiger counter and found radioactivity everywhere: on the door knobs, sink faucets, and even on pencils. The same principle applies to pathogens—gloves must be promptly and properly discarded. In the next chapter, we discuss surveillance that helps guide public policy.

Chapter 8

Surveillance

Summary: National and global surveillance, sometimes conducted over many years, reveals trends in antibiotic resistance that helps guide public policy. Surveillance is based on measurements of MIC with a large number of patient isolates. With a few pathogens, the phenotypic measures of MIC are gradually being replaced with genotypic (DNA) analyses, which require that resistance correlate strongly with particular mutations. Dosing strategies for gonorrhea, animal use of antibiotics, and hospital-to-hospital spread of resistant *Enterococcus* serve as examples where surveillance has revealed a need for changes in antibiotic use strategies.

The previous chapter discussed our immediate response to disease outbreaks and difficulties caused by antibiotic resistance. We now turn to surveillance, one of the long-term, global responses to resistance.

Surveillance Is the First Line of Defense

Surveillance describes the temporal and geographical patterns of antibiotic-resistant disease. Individual patient samples, initially collected to help guide treatment, usually serve as source material. Pathogens in these samples are grown in the laboratory, and the resulting cultures are tested with a variety of antibiotics to determine MIC. Comparison of the MIC of the isolates with breakpoints indicates whether the isolate is susceptible or resistant. As pointed out in the previous chapter, the results are reported back to the attending physician to guide therapy. For patients in hospitals, the results are also collected by infection-control personnel to help identify outbreaks of resistant disease. Government agencies and academic groups then combine local results with those from other hospitals and other regions to provide a status report on resistance for particular geographic areas. This information identifies outbreaks of resistance, guides infection control and public health strategies, and educates the medical community and general public about resistance problems.

The parameter derived from resistance data is called the prevalence of resistance, the number of resistant cases divided by the total number of isolates tested. When the prevalence of resistance is multiplied by the total number of infected patients, the number of persons suffering from drug-resistant disease can be estimated. This is how the World Health Organization projects that 450,000 persons worldwide get MDR tuberculosis each year.

The use of clinical breakpoints to define susceptible and resistant isolates permits data to be condensed into easily communicated ideas. But use of breakpoints gives the word "resistance" an absolute meaning: A pathogen isolate is either resistant or it is not; one resistant bacterial isolate is not categorized as being more resistant than another. In reality, resistant strains do commonly differ in susceptibility, because they differ in the number and type of resistance mutations they carry. Likewise, "susceptible" strains can carry resistance mutations that lower susceptibility to different degrees. This difference between clinical resistance, defined by breakpoints, and the effect of resistance mutations is more than semantic. It influences how we think about resistance and how we design surveillance studies. For example, surveillance data may show that the prevalence of clinical resistance is low, thereby leading to the conclusion that no problem exists. However, many isolates could have lost some, but not all, susceptibility and still be scored as susceptible. This loss of susceptibility, called MIC creep, would be observed if the MIC of the isolates were measured at various times. MIC data obtained over several years make it possible to measure the *rate* (speed) at which MIC is increasing among clinically susceptible isolates. In principle, such detailed longitudinal surveillance using MIC reveals resistance problems early enough to forestall resistance by adjusting antibiotic use and dosing practices. Focusing only on current-year prevalence of resistance can lead to a false sense of security and delayed response.

Surveillance workers also collect data on antibiotic use, often measured as sales of antibiotic or number of prescriptions issued. To compare different geographic regions and human populations, antibiotic use is adjusted (normalized) for the number of persons in the population. That can be taken as persons in a hospital, a country, or even a continent. As emphasized in Chapter 10, "Restricting Antibiotic Use and Optimizing Dosing," antibiotic resistance correlates strongly with use.

The Denominator Effect Lowers Surveillance Accuracy

Most infections are treated empirically, that is, without results of laboratory tests. No patient sample is obtained if the treatment works. These cases, which are absent from the statistics, would otherwise be part of the denominator in the calculation of resistance prevalence (number of resistant cases per total number of cases). The many susceptible cases missing from the denominator cause surveillance to over-estimate resistance problems. This phenomenon is commonly called the denominator effect. The size of the denominator effect can be substantial, as estimated in a study with mild-to-severe food-borne disease conducted in Australia. Unreported cases represented almost 90% of disease caused by *Campylobacter*, *Salmonella*, and shiga-toxin-containing *E. coli*.[181]

Not all pathogens are subject to a denominator effect. For example, tuberculosis surveillance in industrialized countries is not adversely impacted by the failure to report treated cases, because all cases must be reported to health authorities. Nevertheless, not every sample produces testable bacteria. Moreover, immigrants sometimes fear contact with government authorities and do not seek treatment. Corrections are needed for these two factors. MRSA is also a reportable disease in many parts of the United States and Europe. For both tuberculosis and MRSA, obtaining detailed information about resistance patterns is important for public health planning and decision making. For example, when regulatory agencies consider approval of catheters embedded with antibiotics to prevent bacterial colonization by *S. aureus*, the prevalence of resistance to such agents is important to know in advance—if it is low, the antibiotic is more likely to be useful.

Surveillance Consortia Collect and Process Data

Surveillance is performed by a variety of groups (see Box 8-1). In some cases, public agencies, such as the Centers for Disease Control and the World Health Organization, compile data from many local sources. In other cases, small groups of clinical scientists solicit information from hospitals and clinical laboratories, generally with financial support from government agencies. Pharmaceutical companies also participate, because surveillance surveys in the United States and Europe are now a required element of the approval path and follow-up for new antibiotics. Such studies help companies market their compounds after approval if the prevalence of resistance is low. Industry also uses surveys to direct future antibiotic development when the prevalence of resistance is growing.

Box 8-1: Surveillance Networks for Antibiotic Resistance

Many of the major surveillance networks were established in the late 1990s,[182] a time when resistance became recognized as a widespread problem. However, nosocomial infections had been recognized for decades, as evidenced by establishment of the National Healthcare Safety Network (NHSN) in 1970 for U.S. hospitals.[182] By the end of the 1990s, several networks were in place for European hospitals (HELICS, 1994; EARSS, 1998) and U.S. intensive care facilities ((ICARW, 1995). Respiratory infections also received attention in the U.S. (TRUST, 1996) and elsewhere (Alexander Project, 1992; PROTEKT, 1999). Community care centers in Canada and Europe began collating data for resistant urinary infections (ECO-SENS, 1999). As the resistance problem increased in severity, surveillance expanded to cover “common pathogens” in medical centers and outpatient facilities, initially in 30 countries worldwide (SENTRY, 1997). To study the relationship between antibiotic use and resistance, German intensive care units set up a network (SARI) in 2000. These and other surveillance networks establish that antibiotic resistance is a serious, growing problem.

Molecular Methods Provide Rapid Pathogen Identification

Nucleic acid tests (see Box 8-2) enable us to identify pathogens without growing them in the laboratory. The tests are now performed in diagnostic laboratories, but eventually they will be carried out in doctors’ offices. Because diagnostic laboratories currently serve as central data-collecting agencies, shifting the tests to doctors will require expanded reporting to draw test results into surveillance databases. A problem could arise if physicians do not routinely save patient samples after completion of the tests: The samples would not be available to the research community that carries out much of the surveillance work. Thus, adjustments may be needed for surveillance to continue effectively.

Box 8-2: Nucleic Acid-Based Diagnosis

Pathogen species are readily distinguished by nucleotide sequence differences in their ribosomal RNA genes. PCR (see Box A-3) can be performed using primers that amplify regions specific for each of the suspected pathogens. Several methods are available to quickly identify the pathogen species. The type using "molecular beacons" is described below.[34]

A molecular beacon is a short single-stranded DNA molecule (about 25 nucleotides long) in which a four-nucleotide stretch at one end is complementary with a four-nucleotide stretch at the other end. When base pairing occurs, the beacon adopts a stem-loop (lollypop) structure in which the nucleotides in the middle (probe region) remain as a single-stranded loop (see Figure 8.1).

Figure 8-1 Nucleic acid-based detection of resistance. A method using molecular beacons is based on hybridization between a portion of the beacon and the target nucleic acid. Hybridization of the beacon with its target forces apart the fluorophore and the quencher.

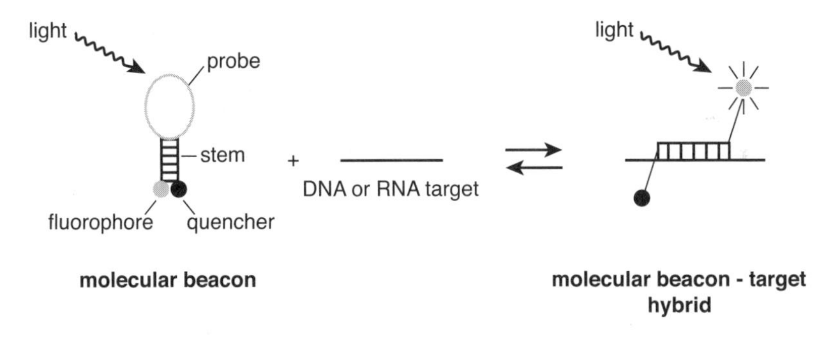

Drawing provided by Salvatore Marras, Public Health Research Institute.

The loop is designed to contain a short nucleotide sequence complementary to the target pathogen DNA. When a molecular beacon is mixed with a sample containing the pathogen DNA, the loop region of the beacon hybridizes (forms complementary base pairs) with the pathogen target DNA sequence. The hybridization reaction forces apart the two ends of the beacon. To measure the separation of the two ends, the beacon is designed to have a molecular structure called a fluorophore attached to one end and a quencher at the other end. The fluorophore emits fluorescent light when stimulated with a particular wavelength of visible light. But when it is next to the quencher, it cannot. Only when hybridization occurs and the two ends of the beacon are forced apart

continues

does the beacon fluoresce. Thus, fluorescence indicates the presence of a particular pathogen. Molecular beacons designed to detect different pathogens can be prepared in which a distinct color is used for each pathogen, thereby creating a multiplex assay in which multiple pathogen tests can be performed simultaneously with a single patient sample.[183]

The MIC is a phenotypic property of an isolate with respect to a particular antibiotic. A mutation is a genotypic property. When a mutation is known to be responsible for resistance, genotypic (nucleic acid) tests can be used to assess resistance. Genotypic tests for resistance are often difficult to interpret because many genes can contribute to resistance and because a gene for a drug target can contain a wide variety of mutations that may or may not contribute to resistance. These unknowns currently limit the use of nucleic acid tests for resistance determination. In contrast, the phenotypic MIC test in which the pathogen is grown in the laboratory takes longer, but the test result reflects the net effect of all mutations for a given drug.

In a few situations, changes in nucleic acid sequence are strongly associated with particular types of resistance. Those situations are likely to be useful for detecting resistance with pathogens that grow slowly and are difficult and/or dangerous to culture. *M. tuberculosis* is one of the best examples (Box 8-3).[35]

Box 8-3: Determining Antibiotic Resistance for M. tuberculosis *by Genotyping*

The *M. tuberculosis* genome is remarkably conserved from one strain to another, and most mutations in genes that encode antibiotic targets contribute to resistance. Consequently, DNA methods have been devised to rapidly determine the resistance profiles of isolates recovered from patients. One strategy combines PCR with nucleic acid hybridization to identify resistance genes. DNA is extracted from the patient sample, and specific regions are amplified by PCR. The resulting mixture of amplicons is then hybridized to a set of DNA probes bound separately to a strip of paper-like material. One probe is to indicate whether the sample contains *M. tuberculosis* DNA, and several others are from sections of genes where mutations are known to cause resistance. Probes for both mutant and susceptible DNA are present for each of the gene sections. If the DNA from the isolate contains a particular resistance mutation, the amplicon from the PCR reaction hybridizes to the region of the strip where the corresponding mutant probe is located. If the amplicon is from resistant DNA, it will not hybridize to the corresponding wild-type probe, which is present at another location on the strip.

After hybridization is performed, the strip is treated so that the hybrids appear as visible bands that are easily seen. For *M. tuberculosis* rifampicin resistance, these DNA assays have an accuracy (sensitivity) of greater than 98% when compared to conventional culture-growth methods.[36, 37] The methods do not work for pyrazinamide resistance because it can arise from a large variety of mutations scattered throughout the target gene. Nucleotide sequence analysis is customary for identifying pyrazinamide-resistant *M. tuberculosis*.

One problem with the PCR-hybridization method is that the tubes used for PCR must be opened to perform hybridization. That can permit amplified DNA (amplicons) to escape and contaminate the laboratory (when the amplicons get in a subsequent PCR, they may distort the result by priming the reaction). Thus, great care must be exercised with "open-tube" assays. "Closed-tube" methods are now being developed. One involves real-time probes that produce fluorescent light when hybridization occurs (see Box 8-2). Mutant DNA can be identified by measuring the temperature at which the hybrids melt (come apart). After hybridization, the temperature in the tube is gradually increased until the fluorescence disappears; melting temperature reflects sequence differences among the two hybridized DNA strands.[184]

An increasing prevalence of resistance will drive efforts to quickly determine both the cause of disease and antibiotic susceptibility profiles for many common diseases. The detailed pathogen information derived from these tests will permit narrow-spectrum antibiotics to be used more often, which will reduce damage to commensal populations and reduce emergence of resistance among commensals. DNA tests will also enable antibiotics to be used in geographic regions where the prevalence of resistance is high: Only patients with susceptible isolates would receive particular antibiotics. Correlation of molecular changes with resistance is an active area of research that is producing promising results. For example, triazole resistance in *Aspergillus* and echinocandin resistance in *Candida* are now known to be limited to specific regions of the *Cyp51A* and *FKS* genes, respectively. Thus, genotype diagnostics may become available for these forms of resistance.

Interpretation of Surveillance Studies

Obtaining data for surveillance can be deceptively simple in concept: Simply collate information from diagnostic laboratories. However, interpreting that data is often not straightforward due to bias in the data collection process. For example, most samples sent to diagnostic laboratories are from hospitals and are thus biased toward patients who are quite ill. Another problem concerns longitudinal studies. The most informative surveillance reveals trends over time, because they enable policy decisions to be made. For such work, one needs the same patient population each year. Problems associated with interpretation and design of surveillance studies are sketched in Box 8-4.

Box 8-4: Questions to Ask of Surveillance

In the 1990s, many surveillance studies were established. After several years, factors that limited the generality of the results began to surface.[185,186] Following are some of the more obvious factors to consider when examining the results of surveillance work:

1. Were only first samples included and how well did the work avoid duplicate samples? Samples from patients failing therapy may be sent many times to a diagnostic laboratory in attempts to find an effective treatment. Later, samples may reflect isolates recovered after being challenged with antibiotic, and they may not be representative of the overall population.
2. Was bias created by the patient population? Patients in intensive care units (ICUs), non-ICUs, and out-patients differ in disease severity and resistance prevalence.
3. When data were derived from a variety of diagnostic laboratories, was the information obtained using the same standardized method? Were the patient populations carefully defined and the same for the various source laboratories?
4. In longitudinal studies, were the patient populations comparable for all years of the study?
5. Was the denominator effect evaluated and included in the final conclusions?
6. When samples are from diagnostic laboratories, particularly those in hospitals, are the conclusions relevant to the community; 75% of human antibiotic use is in the community, yet surveillance data are largely from hospitals.[185]
7. Is there possible bias due to the sources of financial support for the study? Surveillance work is expensive, often funded by industry, and sometimes used in marketing.

Surveillance Indicates Resistance Problems with Gonorrhea

Surveillance work is revealing a problem with gonorrhea that may require a change in dosing policy. In the 1960s, gonorrhea was easily treated with a single dose of penicillin, but resistance eventually developed. Strains resistant to tetracycline and penicillin spread from the Philippines to other areas of the Pacific Basin, including Hawaii. From there, they traveled to California, then to the rest of the United States, and eventually to Europe. Fluoroquinolones were substituted for penicillin, and for several years these compounds were used successfully as single-dose cures. However, by 2006, resistance had seriously eroded the fluoroquinolone treatment option. Potent relatives of penicillin (cephalosporins) are now used, but they are not likely to provide a long-term solution.[22]

The organism that causes gonorrhea is inherently susceptible to many antibiotics. Single-dose treatments were deemed desirable because the disease is common among sex workers who sometimes have difficulty adhering to a treatment schedule. But curing symptoms and removing forms of infection that are easily detectable does not prevent the emergence and spread of resistance. Continuing with the current, single-dose treatment policy may eliminate antibiotic options for gonorrhea, leaving us with only low-tech preventive solutions, such as condom use.

Policy Changes Are Occurring in Agricultural Practice

Agricultural practices foster the development of antibiotic-resistant intestinal pathogens. These microbes, mostly *Salmonella* and *Campylobacte*r, move from farms to homes and restaurants. Because good sanitation systems stop the bacteria from spreading beyond the immediate consumer, most of the preventive effort is to block contaminated food from reaching agricultural markets. Surveillance linked fluoroquinolone resistance to agricultural use of the compounds (see Box 8-5). To reduce the problem of resistance, government approval for fluoroquinolone use with poultry was withdrawn.

Box 8-5: Removal of Fluoroquinolones from U.S. Poultry Use

Campylobacter infections usually cause mild cases of diarrhea that are self-limiting. However, they are occasionally serious. Risk factors for infection include eating and handling foods of animal origin, particularly poultry. In 1995, the FDA allowed fluoroquinolones to be used for respiratory diseases of poultry. The guidelines permitted addition of the drugs to drinking water used by chickens, which meant that 10,000 to 30,000 birds (an entire chicken house) were treated at once. Examination of *Campylobacter* isolates taken from humans in the early 1990s, before fluoroquinolone use with poultry, revealed none to be fluoroquinolone resistant. By 2001, 41% were resistant. Estimates for the U.S. in 1999 indicated that 157,000 persons had become infected with fluoroquinolone-resistant *Campylobacter*. In about 9,000 of those cases, patients were treated with a fluoroquinolone even though the infections were already resistant. (Knowledge of pathogen susceptibility lags behind initial treatment.) In another survey, 44% of 180 retail chicken products were contaminated with *Campylobacter*; 10% were fluoroquinolone resistant.

As a result of these and other studies, the FDA issued a notice in 2000 that it considered fluoroquinolone-resistant *Campylobacter* to be a threat to human health. At the time, two companies supplied the agricultural market with fluoroquinolones. One withdrew its compound from the market, whereas the other fought the ruling in court. The FDA side eventually prevailed, and in September, 2005, fluoroquinolones were removed from the U.S. poultry market.[187]

Studies with food animals are particularly suited for examining relationships between antibiotic usage and resistance (see Box 8-6). One reason is that usage levels are more easily controlled with food animals than with humans. Another reason is that by collecting samples from slaughterhouses it is possible to obtain data that are representative of the overall healthy food animal population.

Box 8-6: Surveillance in Danish Food Animals

In Denmark, where intensive food animal production is practiced, antibiotic use and resistance has been extensively studied with cattle, pigs, and chickens as part of a program called DANMAP.[135] Three bacterial categories were monitored on a monthly basis. One concerned animal pathogens (*Staphylococcus hyicus* and pathogenic *E. coli* strains) collected from diagnostic laboratories that service veterinarians. Another focused on zoonotic bacteria that cause disease in humans (*Salmonetla enterica, Campylobacter* species, and *Yersinia enterocolitica*). These organisms were collected from animals at slaughterhouses and therefore focused on healthy animals. A third group included indicator bacteria (*E. coli* and *Enterococcus faecium/faecalis*) that were also obtained from slaughterhouses. Samples were limited to one per herd, flock, or farm to eliminate duplicates. For each isolate, the MIC was measured for a variety of antibiotics, and a sample was retained in a frozen collection for future analyses. Investigators also determined the level of usage for the relevant antibiotics.

Because samples were taken before and after dramatic restrictions on use of growth promoters, the investigators had an excellent opportunity to examine relationships between use and resistance. Data with vancomycin-resistant enterococci (VRE) were particularly interesting. Use of avoparcin, a vancomycin-like compound, as a growth promoter was banned in 1995, and within a year, the recovery of VRE from chickens dropped dramatically. However, little change was seen with pigs. Three points were relevant:

1. VRE isolates from pigs were also resistant to macrolides.
2. The genes responsible for vancomycin and macrolide resistance were on the same mobile genetic element.
3. Macrolides continued to be used with pigs.

A few years later, when macrolide use dropped dramatically, the prevalence of VRE from pigs also dropped. These data argue that with the drugs and bacteria examined, the prevalence of resistance correlates strongly with antibiotic usage. The data also indicate that resistance can have a significant fitness cost, because resistant bacteria were replaced by susceptible ones.

Removal of avoparcin from use (refer to Box 8-6) came too late: Antibiotic-resistant *Enterococcus faecium* moved from the farm to hospitals where it became a global problem (see Box 8-7). *E. faecium* now threatens to transfer vancomycin resistance to MRSA in hospitals, which would severely limit our antibiotic options with *S. aureus*.

Box 8-7: Global Spread of Enterococcus

Enterococci are normal inhabitants of human and animal digestive tracts, but when they get in the blood stream, they cause serious disease. Two species of primary concern are *Enterococcus faecalis* and *E. faecium*. Recently *E. faecium* became an important pathogen in hospitals due to its acquisition of resistance to antibiotics such as ampicillin and vancomycin. Isolates carrying the *vanA* (vancomycin-resistance) transposon were found in Europe in 1987, and within 10 years vancomycin-resistant Enterococci (VRE) were associated with more than 25% of bloodstream infections in U.S. hospitals. Prior to 1997, a vancomycin-like antibiotic called avoparcin had been used in Europe as an animal growth promoter, and in Europe VRE are frequently found in farm animals.

In contrast, avoparcin and related glycopeptides were never approved for use in U.S. agriculture, yet VRE spread to U.S. hospitals. To understand the apparent spread of VRE, DNA was obtained from a variety of isolates and examined.[188] A bacterial "family tree" was constructed to relate the isolates, and from this analysis a subfamily called complex-17 was identified as being common. (142 of 411 isolates were members of the subfamily.) Complex-17 was rare among isolates from animals (1 of 96) and humans in the community (3 of 57). But hospital surveys turned up many (15 of 64), and among human disease isolates complex-17 was common (95 of 162). In an outbreak of VRE, 28 of 32 isolates were members of the complex-17 subfamily, the result expected from clonal spread.

Complex-17 members have adapted to the hospital environment and are now globally distributed. Precisely how that occurred is unknown. Also unknown is which genes contribute to the "hospitalization" of complex-17. Almost all the complex-17 isolates are ampicillin resistant, but only half carry a pathogenicity island, a stretch of "foreign" DNA that contains virulence genes. Thus, the group appears to have acquired ampicillin resistance before the pathogenicity island. Because vancomycin resistance is widespread,[188] it may have been acquired in multiple strains from repeated exposure to avoparcin on farms.

Perspective

Surveillance tells us when our antibiotic use policies fail to control resistance. In general, the global distribution of antibiotic resistance is uneven: Pockets of resistance develop and expand. In many cases, we have good explanations for the development of resistance, and in some situations the resistance problems were predictable. For example, the MDR tuberculosis outbreak in the prisons of the former Soviet Union was certain to spread when inmates were released to the community without adequate support.[155] But providing workable solutions is difficult due to the scale of the problem and to the failure of the general public to heed the warnings provided by surveillance. Some pockets of success are also seen, such as the control of hospital-associated MRSA in Denmark and The Netherlands. But most of the hope is placed on the development of new antibiotics, which we discuss in the next chapter.

Chapter 9

Making New Antibiotics

Summary: When resistance erodes the usefulness of antibiotics, we develop new ones, using natural products and large collections of synthetic compounds. Robots enable us to test millions of compounds for activity (high-throughput screening). The key to these methods is the availability of a simple, relevant test (assay), often a biochemical surrogate for a process occurring in the living microbe or in virus-infected host cells. Unfortunately, resistance has created a fundamental problem for drug discovery efforts: Use of new compounds is restricted to protect from resistance, thereby reducing financial incentive and diminishing antibiotic discovery efforts.

The examples described in the previous chapter illustrate how surveillance warns us when resistance begins to erode the utility of existing agents. Surveillance also alerts the pharmaceutical industry to possible opportunities for new products. This chapter focuses on how new antibiotics are found.

New Antibiotics Are Temporary Solutions

Use of any of today's antibiotics will ultimately result in the selection of resistant mutants; consequently, new antibiotics are only likely to delay, not eliminate a resistance problem. For example, when methicillin replaced penicillin with *S. aureus* in the 1960s, it gave great hope. But now we have a form of drug-resistant *S. aureus* that the public knows only by an acronym (MRSA) representing methicillin resistance. In another example, the fluoroquinolone moxifloxacin is being tested for addition to our antituberculosis arsenal. Moxifloxacin has a much longer life span (half-life) in patients than many other antibiotics commonly used for tuberculosis. Consequently, moxifloxacin, when used in combination therapy, will continue to be present after the other agents have been metabolized or excreted. Although moxifloxacin may improve cure rate, in the long run we expect it to produce widespread fluoroquinolone resistance, because for much of the dosing interval moxifloxacin will be present and acting as monotherapy. (No antibiotic is deliberately used as monotherapy with *M. tuberculosis* because resistance emerges so readily.) Thus, new antibiotics will be needed. In the remainder of the chapter, we discuss how they are found.

Model Systems Are Used to Speed Drug Discovery

Some pathogens are too difficult or dangerous to handle on a routine basis. Consequently, preliminary experiments are frequently performed with similar, but more easily studied, relatives. *M. tuberculosis* serves as an example. This pathogen can be cultured using synthetic medium, but almost a month of incubation is required before colonies are visible on agar. (Only a day is needed with *E. coli.*) Moreover, *M. tuberculosis* is dangerous, especially when it is already resistant to several antibiotics. Work must be performed under biosafety level 3 containment, which means that workers must wear personal protection equipment (for example, gowns, respirators, shoe covers, and gloves). The work is carried out inside a specialized laboratory under negative air pressure, and the organism is handled inside a biosafety cabinet that has contained airflow. Because laboratory workers must be specially trained, management flexibility is limited. Moreover, each new experimental protocol must be approved by safety committees. Safety issues can be reduced by working with *M. bovis* BCG, a relatively safe strain of mycobacterium used to prepare vaccines. *M. bovis* also grows slowly, and like *M. tuberculosis*, it can be induced to undergo growth arrest, a feature likely to reflect the capability of *M. tuberculosis* to enter a dormant state after infection. A more distantly related model organism is *M. smegmatis.* This fast-growing soil bacterium shares many properties with *M. tuberculosis*, and it is easier to handle than *M. bovis* BCG. Even *E. coli* is a good model organism for some fluoroquinolone work because quinolone action is similar in all bacteria. However, *E. coli* would be a poor choice for studies of cell walls due to the special nature of mycobacterial walls.

Bacillus anthracis provides another example. This bacterium grows rapidly, but it is extremely dangerous because it forms spores that can easily enter human lungs and germinate. The growing bacteria cause rapid, acute pneumonia in addition to producing a lethal toxin. In its virulent form, *B. anthracis* is studied only in registered laboratories under carefully guarded conditions. Consequently, preliminary work is often carried out with a strain lacking a plasmid required for virulence. With this pathogen, most virulence factors are probably unrelated to antibiotic susceptibility, which makes plasmid-free strains suitable for preliminary drug studies. The key idea is that the model organism should behave much like the pathogen with respect to the particular property of interest.

When principles are clear from work with model organisms, confirming experiments are carried out with the pathogen. Such confirmatory work is important because model microbes may differ from pathogens in ways that are difficult to predict. In the case of *M. tuberculosis*, the National Institutes of Health has contracted with universities and other institutions to maintain testing facilities where research groups can submit new compounds for validation of results obtained with model organisms.

Natural Products Are a Source of Antibiotics

During the competition for ecological niches, many life forms produce compounds that poison other organisms. (Even bacteria that normally reside in the human gastrointestinal tract produce compounds that are toxic to a variety of pathogens.) At low concentrations, these toxins may serve as signaling molecules, whereas at high concentrations they may poison competitors. Some of these compounds have been our antibiotics since ancient times (see Box 9-1). Today, the natural poisons identified from plants, animals, fungi, and bacteria serve as chemical starting points for making even better antibiotics.

Several general approaches are followed for finding new antibiotics in natural sources. One is to screen aqueous and solvent extracts of diverse natural sources for compounds that either stop growth or kill the pathogen in question. For example, if one wanted to find a new chemical to block growth of *Salmonella*, extracts would first be prepared from rich sources of natural products, such as from cyanobacteria, ocean sponges, or the soil microbe *Streptomyces*. Those extracts would then be added to a broth medium in which *Salmonella* grows. *Salmonella* would also be added, the culture would be incubated, and then cultures would be examined to find those in which no growth occurred. Follow-up chemistry would be used to purify active compounds, to determine their structure, and to make more effective derivatives. This screening process is called bioassay-guided isolation of natural products; it is the most common approach used to find natural products active against pathogens. In some cases, the original compound from the natural source is subsequently prepared synthetically for study and ultimately for use as a drug. In other cases, obtaining the agent directly from plant or bacterial cultures may be the most economical production strategy. Large vats are used to grow antibiotic-producing microbes, and methods have been developed to extract the active ingredient economically.[123]

Box 9-1: Natural Sources of Antibiotics

Both the ancient and modern pharmaceutical worlds have relied on natural products to develop important drugs. Among these are the drugs digitalis from foxglove and morphine from poppies. The Greek physician Hippocrates, the father of Western medicine, noted in the Fifth Century BC that a bitter powder (salicylic acid), extracted from willow bark, could reduce fevers and lessen aches and pains. Traditional Chinese healers have relied on natural herbal antibiotics for more than 4,000 years. From 2698 to 2596 B.C, the Inner Canon of the Yellow Emperor was written as the seminal medical text of ancient China, establishing the foundations of traditional medicine. During the Tang Dynasty, the volume *Materia medica* (657 AD) described 833 medicinal substances taken from minerals, metals, plants, herbs, animals, vegetables, fruits, and cereal crops. Other classical Chinese medical works, such as *Theory of Febrile Diseases* and *Synopsis of the Golden Cabinet* by Zhang Zhongjing (150–219 AD) and *Treatise on Differentiation and Treatment of Seasonal Febrile Diseases* by Wu Jutong (1798), have been used to treat a variety of infectious diseases.

One of the richest natural sources of antibiotics is the bacterial genus *Streptomyces*. These soil organisms make a variety of compounds that inhibit the growth of bacteria and fungi. Indeed, *Streptomyces* species (see Figure 9-1) have contributed more than two-thirds of existing (natural) antibiotics including nystatin (from *S. noursei*), amphotericin B (from *S. nodosus*), erythromycin (from *S. erythreus*), neomycin (from *S. fradiae*), streptomycin (from *S. griseus*), tetracycline (from *S. rimosus*), vancomycin (from *S. orientalis*), rifamycin (from *S. mediterranei*), chloramphenicol (from *S. venezuelae*), and puromycin (from *S. alboniger*). New techniques extend the limits of *Streptomyces* metabolism by altering culture conditions to yield an even richer chemical milieu for compound screening.

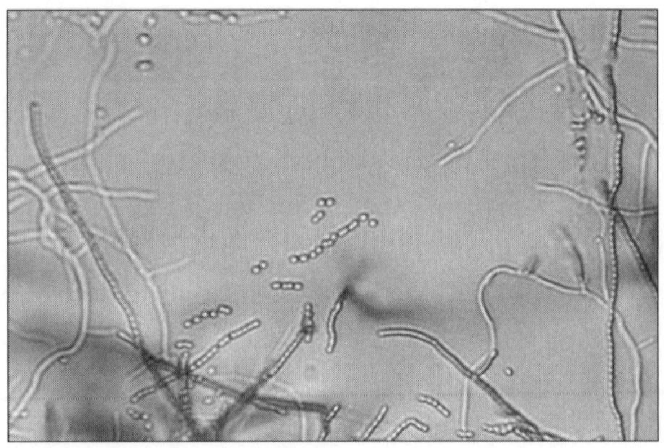

Figure 9-1 *Streptomyces* sp. Photomicrograph of *Streptomyces* growing on agar. Long chains of spores are shown among branching filaments called mycelia.

Public Health Image Library #2983.

High-Throughput Screening Accelerates Antibiotic Discovery

Microbes have evolved defenses against the toxic effects of antibiotics produced by other microbes and against the antibiotics they themselves make. The genes encoding these defense systems can move among pathogen populations and eventually erode the medical usefulness of natural antibiotics. Indeed, any antibiotic derived from a natural source, or based on a natural antibiotic's mechanism of action, is an antibiotic for which resistance is already present in nature. In principle, synthetic antibiotics, particularly those for which the mechanism of action differs from that of natural antibiotics, should be less subject to natural forms of resistance. Consequently, methods have been developed to find new antibiotics that act against new targets. For example, new agents are sought that would act on enzymes or receptors not targeted by natural agents, or the new agents would act on an old target but in a new way. One search strategy is to create large collections (libraries) of drug-like chemicals. Robotic systems are then used to quickly test compounds in the libraries for antibacterial activity (commonly hundreds of thousands of different compounds). When active compounds are found, subsequent work determines how they act on pathogens and whether the compounds are good prospects as drugs. Alternatively, if the nucleotide sequence of the genome of a pathogen is known, genes and gene products required for survival can be identified. For example, a gene required for survival may be found to encode an enzyme involved in making bacterial cell wall. Libraries of compounds can then be screened for the ability to inactivate the gene product.

Starter compounds emerging from the initial screens, often termed lead or pre-lead compounds, may not easily get inside pathogen cells; consequently, tests with living organisms can fail to reveal agents that are highly potent against the cellular target. To bypass the drug uptake problem, biochemical assays are designed in which purified components (enzymes, receptors, transporters, and so on) are used as drug targets. These assays help identify compounds that act on particular processes that are unique to particular pathogens, thereby reducing the chance for toxicity problems. In general, biochemical assays are set up to find inhibitors of key enzymes required in critical biosynthetic processes of a given pathogen. Before developing such tests, one would like to know that inactivating the target enzyme seriously harms the pathogen. Harm can often be established by showing that mutations in the gene encoding the potential protein target cause cell death, or at least cessation of growth. In some cases, logic is sufficient. For example, HIV is unlikely to set up a productive infection if it cannot replicate its genome. Consequently, an obvious first step to controlling this virus was to find inhibitors of HIV reverse transcriptase, the enzyme that converts the genetic information of the virus from an RNA form to the DNA molecule that inserts into human chromosomes.

When a critical enzyme target is identified, the next step is to create a biochemical test for a reaction that requires the target protein. For example, if we were looking for an inhibitor of DNA replication, we would set up a measure of replication, perhaps involving the incorporation of small nucleotides into large DNA. A critical feature is that the assay needs to be easy for robots to perform quickly with large numbers of compounds. After a test for enzyme activity is in place, libraries of compounds are tested for the ability to block that activity. The overall process is called high-throughput screening. Compounds found to block enzyme activity are then modified chemically to be more active, both with biochemical assays and with tests on living pathogens. The desired compounds must also exhibit low toxicity to human cells. Indeed, most new antibiotics that fail to reach clinical use do so because they lack adequate or appropriate ADMET (Absorption, Distribution, Metabolism, Excretion, Toxicity) profiles. Thus, the task is to identify a good drug target at the beginning of the process and then find lead molecules that can be chemically modified to fine-tune or alter ADMET properties without losing potent antibacterial activity.

Rational Drug Design Can Identify Antibiotics

Antibiotic discovery has historically involved searching through large numbers of compounds for a few with defined activity. Leads are then followed by rationally modifying compound structure to design a better drug. However, a computational, structure-based approach has also emerged. In this strategy, which is an aspect of drug discovery known as computer-assisted drug design, the three-dimensional structure of a target protein is determined at such a fine level that the position of every atom is known. Obtaining the three-dimensional structure of a protein often derives from a technique called X-ray crystallography. The method involves obtaining crystals of the molecule of interest. (An example of a crystal can be seen by allowing a concentrated salt solution to stand for a long time; eventually, a large crystal of salt forms in the bottom of the jar holding the solution.) The same process occurs with proteins if the correct conditions are found. Although finding those conditions requires considerable trial-and-error experimentation, many successes have been achieved. When a protein crystal is available, it is used as a target for X-rays.

By examining how the X-rays bounce off the target relative to known reference points, the position of every atom in the test molecule can be deduced.

When the structure of a protein is known, it is often possible to identify parts of the protein crucial for its activity. The structure can then be entered into a computer program along with structures of potential inhibitors. The computer program then attempts to fit (dock) the "inhibitor" with binding sites on the target protein. This *in silico* or computational strategy to drug design enables chemists to examine millions of compounds for protein binding before beginning laboratory-based work. Compounds calculated to be best at binding the protein active site are then synthesized, examined for inhibition of the protein's function, and tested for activity against cultured pathogens. When active compounds are found, docking these compounds onto the protein structure provides a picture of drug-protein binding that can be used to rationally design next-generation compounds. This type of strategy led to inhibitors of HIV protease. As with high-throughput screening, structure- and computer-assisted drug design depends on first establishing a suitable target protein. Such proteins are often termed "druggable," which indicates that inhibiting the protein by a small molecule is a likely therapeutic option.

New Antibiotics Must Have Few Side Effects

The medical community seeks to "do no harm" with its medicines. Consequently, new compounds undergo extensive testing before coming to market. Although activity against the target is required, ADMET properties (previously discussed) are extensively evaluated using in vitro assays and models. Most lead molecules that fail to become drug candidates do not fail because they lack antimicrobial activity; they fail because they lack appropriate physical properties for dosing, for distribution throughout the body, and for desired breakdown and excretion.

An evaluation of toxicity with cultured human cells is one of the first tests conducted with a new lead compound. Healthy living cells are able to keep certain dyes outside the cell; if a cell is damaged, the dyes get inside. New compounds can be tested to determine whether they eliminate the ability of human cells to exclude dyes. Agents that pass the initial in vitro tests are then examined with laboratory animals, such as mice. Then the compounds are tested for safety and efficacy in small clinical trials with human volunteers. Data from these experiments are presented to regulatory agencies to expand clinical trials and ultimately to gain approval to sell the compound.

Even after a compound reaches market, its safety record continues to be monitored by its pharmaceutical suppliers and by governmental agencies such as the Food and Drug Administration. This post-market monitoring is important to identify rare side effects and particularly dangerous uses that might not be apparent from the limited studies performed prior to government approval. Post-market antibiotic toxicity problems may become apparent only after large numbers of patients receive the drug.[77]

Antibiotic Discovery Faces a Fundamental Economic Problem

If a company develops a highly effective antibiotic, the medical community tends to limit its clinical use to minimize the development of resistance. That forces the pharmaceutical company to charge a high price to recoup investment costs. High prices are politically and socially unpopular because the public has grown accustomed to having access to safe, highly effective, and inexpensive antibiotics. (As previously pointed out, antibiotics are sometimes given away free by grocery stores.) Moreover, an effective antibiotic will cure disease and ultimately no longer be needed by the patient. Consequently, antibiotic development in its traditional form is often not feasible as a for-profit enterprise. At the same time, the companies see excellent markets for other types of medicines, especially those used for chronic diseases. With these diseases the drugs must be consumed every day for the remainder of the patients' lives. These considerations explain why microbiology and antibiotic discovery divisions have disappeared from large pharmaceutical companies. Small biotech companies have started to identify new antibiotic targets, but they face the same fundamental problem: Developing a new antibiotic might cost more than can be recouped before patents expire and competition from generic suppliers begins.

Because pharmaceutical companies, or any company for that matter, cannot survive by developing and producing a product that they know will ultimately lose money, development costs must be reduced, or governments must heavily subsidize the effort. Alternatively, companies could seek new anti-infectives that will severely restrict the emergence of resistance (see Box 10-8): Extending the time before resistance develops should increase the likelihood that development costs will be recovered because sales could be high. Such an alternative has yet to become a focus of drug discovery efforts. We expect the decline in antibiotic discovery to continue until new ways are found to finance antibiotic development or to increase antibiotic longevity.

Perspective

For decades, we have used a two-pronged approach to hold back antibiotic resistance: Surveillance tells us when we are losing control over resistance, and new compounds help us regain it. With each cycle the public health official breathes a sigh of relief; the physicians' confidence in the pharmaceutical industry is vindicated; patients praise the latest miracle drug; and the pharmaceutical company gets another round of patent protection. This cycle exists because we use antibiotics. Stopping the cycle would cause medicine to revert to a pre-antibiotic era of infectious disease. Thus, we consider the cycle necessary, but sub-optimal. Several issues are now likely to change the cycle. First, the pharmaceutical industry is dropping out of antibiotic discovery for the financial reasons discussed. Second, government support for research is often directed more at quick fixes (for example, providing dollars for a company to move a new compound through trials) than at supporting the new knowledge that will produce the next generation of break-through discoveries. Third, tests for pathogens are shifting toward antibody and nucleic acid assays that do not require growth of the pathogen in the laboratory. If samples are not routinely saved, part of our ability to see the emergence of resistance will be lost. Collectively these changes compromise our ability to *respond* to resistance. We must now shift away from a response strategy to one in which we preempt resistance; we must stop resistance early by halting the amplification of resistant mutant subpopulations. In the next chapter, we discuss two ways to accomplish this.

Chapter 10

Restricting Antibiotic Use and Optimizing Dosing

Summary: Antibiotic resistance begins with the selective amplification of a small number of resistant mutants following administration of agents that permit mutants, but not wild-type (susceptible) cells, to reproduce. In some cases, antibiotics also stimulate the creation of mutants. Several types of action are expected to slow these processes. One is to reduce antibiotic consumption by 1) encouraging consumers and medical personnel to use antibiotics only when they are highly likely to be effective and 2) reducing agricultural use of antibiotics. Another is to improve waste disposal to halt environmental contamination and creation of resistance genes among environmental microbes. A third action is to make dosing regimens more stringent so that mutant subpopulations of pathogens do not selectively amplify. The mutant selection window hypothesis provides a framework for deciding how stringent dosing regimens must be. A fourth is to deliberately seek new compounds that restrict resistant mutant growth.

Every time we use an antibiotic we apply selective pressure to microbial populations. Hundreds of millions of doses, taken over decades, are beginning to take their toll: Large surveillance studies establish that the resistance problem has been serious for more than a decade and is getting worse. (U.S. hospitalizations with resistant infections doubled from 2000 through 2005.[6]) In this chapter, we consider changes in antibiotic consumption philosophy that could slow the emergence of resistance.

Antibiotic Conservation: Use Less Often When Unnecessary and Higher Amounts When Needed

Development of resistance progresses through three stages. First, resistant mutants are created in a pathogen population by mutation and by horizontal movement of resistance genes from other microbes and viruses. At present, we can do little about either of these natural processes. The second step is gradual enrichment and amplification of the small number of resistant mutants in the pathogen population. This is the step that *must* be blocked if we are to control resistance. The third stage, discussed in Chapter 7, "Transmission of Resistant Disease," is the dissemination of resistant strains. This is where we currently put most of our effort. Unfortunately, controlling the spread of resistant outbreaks is largely a holding action. We must also focus on stopping the emergence of resistance.

Emergence of resistance is influenced by two major factors, the amount of antibiotic we use and how we use it. Overall usage encompasses a variety of factors such as inappropriate use, food animal use, and environmental contamination. We clearly use too much antibiotic when these aspects are considered. How we use antibiotics (design of dosing regimens) is currently determined by efficacy (cure) and safety, not by the capability of a protocol to restrict resistant pathogen growth. New dosing ideas to restrict the emergence of resistance are framed by the mutant selection window hypothesis, which leads to the conclusion that we often use too little antibiotic when treating. Higher doses are needed to stop mutant growth and conserve antibiotic efficacy. Thus, we currently use antibiotics when we shouldn't, thereby overusing them. But when we need them, we tend to use too little by dosing too low. We first consider using antibiotics less often.

Human Consumption of Antibiotics Correlates with Resistance

Little doubt exists that antibiotic use is the cause of resistance. Pathogen isolates obtained before the discovery of a given antibiotic class are all susceptible, and subsequent antibiotic consumption correlates with the prevalence of resistance. Several large studies have established the correlation. One occurred in the late 1980s when healthcare officials in Finland noticed that the prevalence of erythromycin resistance in *Streptococcus pyogenes*, a cause of serious infections, rose from 5% (1988–89) to 13% (1990). Consumption of erythromycin in Finland had almost tripled a few years before the increase in resistance. The Finnish government ordered the use of erythromycin to be severely reduced in outpatient settings. In 1991 and 1992 the consumption of erythromycin and other macrolides was cut in half, and this low level of use was maintained. For a year, the prevalence of resistance increased (to 19% in 1993), and then it began to fall, reaching 8.6% in 1996.[189] In this example, resistance prevalence responded to reduced consumption after a delay of several years.

In another example, penicillin use and prevalence of resistant pneumonia were tabulated for European countries. The relationship between antibiotic consumption and the prevalence of resistance was clear (see Box 10-1). These and similar observations from other work[135, 190] encouraged the medical community to seek reduced consumption of antibiotics.

Box 10-1: Antibiotic Use and Resistance

The prevalence of resistance is related to use: Greater use correlates with higher prevalence. An example is seen when antibiotic consumption in European countries is plotted against the prevalence of resistance in each country (see Figure 10-1).

Figure 10-1 Relationship between penicillin consumption and penicillin resistance in European countries. Outpatient sales are measured as defined daily dose (DDD).

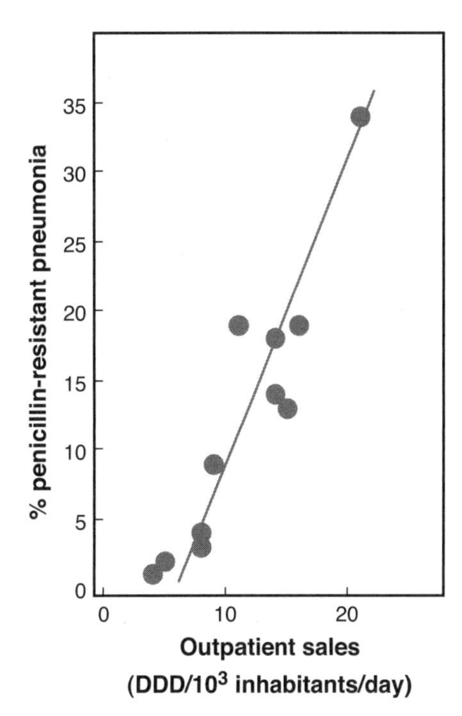

Figure redrawn from data in Bronzwaer, S.L., Cars, O., Buchholz, U., et al. "A European Study on the Relationship Between Antimicrobial Use and Antimicrobial Resistance." *Emerging Infectious Diseases* 2002; 8:278–282.

For pneumococcal infections, the prevalence of penicillin resistance is 25–50% in France, Spain, and Romania, whereas it is much lower in the UK and Holland. Studies show that the average French person consumes three times more antibiotic than a person in Holland. The prevalence of antibiotic resistance is traditionally high in Spain where antibiotics are easy to buy over-the-counter.

continues

Although access to antibiotics appears to be part of the reason that use is higher in some countries, cultural factors are also important, as indicated by a study of two nearby cities, one in Holland and one in Belgium. When both were surveyed for their approach to illness, the Dutch expressed the attitude that their bodies would clear infection. In contrast, the Belgians focused on possible complications of disease and wanted the pathogens eradicated. The Dutch had much lower antibiotic use and lower prevalence of resistance.[191] Such data indicate that simply legislating tighter regulations in Belgium might not solve the problem.

Limiting Human Consumption of Antibiotics

One way to lower consumption is to increase restrictions on use, particularly when the type of use is inappropriate. Almost half of antibacterial consumption in hospitals is considered by some researchers to be inappropriate.[192] In the community, treatment of the common cold serves as a good example of inappropriate use. Antibacterial agents fail to cure this illness because it is caused by a virus. Nevertheless, patients often insist that physicians prescribe antibacterials for colds. Patients feel that they should get something tangible when they visit their physician, and no effective antiviral agent is available for colds. Moreover, patients and many physicians believe that taking antibiotics is generally innocuous. The pressure is so intense that some physicians admit that they would rather write prescriptions than argue with patients. In many cases, the patient will simply go elsewhere to get a prescription if the physician does not comply. The situation is worse when these agents are sold without requiring a prescription. As pointed out, the prevalence of resistance is especially high in countries, such as Spain, where patient-driven dosing prevails. A comparison of France and Germany illustrates some of the issues likely to be important for efforts to reduce antibiotic consumption in the community (see Box 10-2).

Box 10-2: Antibiotic Use in France and Germany

A comparison of two adjacent countries with similar levels of healthcare reveals striking differences, some of which may account for France having a prevalence of antibiotic-resistant pneumococci three times that of Germany.[193] From 1985 to 1997, France had almost three times more antibiotic sales, and during that time Germany tended to use narrow-spectrum agents more than France. For common colds, 50% of consultations with physicians resulted in antibiotic prescriptions in France, whereas the number was only 8% in Germany. German physicians tended to take a wait-and-see attitude. For respiratory tract infections, German doctors ordered diagnostic tests 40% of the time compared with 20% in France. Germans also received higher doses of penicillin than the French, especially with children.

Other, less quantifiable differences have also been pointed out.[193] For example, French persons seek antibiotics for a cough and sputum production more often than Germans, who consider antibiotics to be overmedication. Consequently, patients don't press for antibiotics as often in Germany as in France. Another factor may be the practice of sending young children to preschool where many communicable diseases are spread. Such is the case for most French children. For a comparable age group, the number is only 10% in Germany. Still another factor may involve breast-feeding, because some immunity is transferred in breast milk. Breast feeding is thought to occur more often in Germany.

Regulatory practices may also contribute to differences. Antibiotics have been more expensive in Germany, which may have pushed the Germans toward use of cheaper, narrow-spectrum agents. Moreover, pharmacies in France made greater profits on newer, more expensive antibiotics, whereas in Germany the pharmacist's share was higher with less expensive agents. Thus, many factors appear to contribute to the higher prevalence of resistance in France.

Another problem is the patient who takes antibiotics only until disease symptoms disappear or who stops early due to minor side effects. This is called nonadherence to therapy. If the infecting pathogen population is not reduced to low enough levels, it regrows and creates a relapse of disease. Then another round of antibiotic treatment is needed. With some disease-antibiotic combinations, emergence of resistance occurs more often when patients have received a prior treatment, even when that treatment appears to have cured infection. Creating a need for repeated cycles of antibiotic treatment is unwise. As expected, surveillance work shows a correlation between nonadherence and the prevalence of resistance.[182]

Other types of self-medication are also common. One is seen with nonprescription antifungal treatment of vaginal yeast infections (see Box 10-3), which occur in nearly all adult women. In this situation, resistance is silent until onset of more serious illness. Another is the practice of patients saving left-over prescriptions for later use. In some countries, almost half the human antibiotic use falls into the save-for-later category. Using old antibiotics is a bad idea because oxidation tends to reduce potency. Surveys reveal that self-medication is often based on previously successful self-medication. That is understandable because resistance has not been a common problem on an individual patient basis. However, as the prevalence of resistance increases, the inability of patients to select optimal compounds and doses will make self-medication increasingly risky.

Box 10-3: Vaginal Yeast Infections

The female genitourinary tract is colonized with yeasts, in particular *Candida albicans*. These fungi are normally balanced by bacterial flora. The use of antibiotics or changes in the local environment due to hormones or other physiological conditions can result in proliferation of yeasts. As a society, we encourage individuals to self-diagnose and self-treat by making available over-the-counter antifungal agents that are applied topically to the surface of mucosal membranes. Although individuals with such infections are advised to seek medical care and follow instructions carefully, compliance is a major problem.

In 2001, nearly 2% of yeast samples cultured from the vaginal tracts of women using over-the-counter products were drug resistant.[194] Although this overall resistance prevalence seems low, the total *number* of women colonized with drug-resistant yeasts is high. Yeast infections are generally not life threatening, but during serious illness, these colonizing strains may become infecting strains that can cause life-threatening invasive fungal disease. If the strains are already resistant, antifungal therapy will do little good. Moreover, self-diagnosis fails to identify the initial cause of vaginitis; consequently, cases caused by bacteria will be treated with an ineffective agent.

Agricultural Use Contributes to Antibiotic Consumption

Farmers use roughly ten times the antibiotic tonnage as employed in human medicine. Some of this antibiotic use is for sick animals, some is for prophylaxis, and some is to improve the growth and performance of animals.[195] Farmers found, in the 1940s, that animals grow faster when fed dried parts of *Streptomyces aureofaciens*, which contains a derivative of tetracycline.[196] In 1951, the United States Food and Drug Administration approved the use of antibiotics as growth promoters, and European countries followed with similar approvals. Farmers' groups have argued that concentrations of these growth promoters are so low that they have little consequence for resistance. How antibiotic growth promoters work is not known precisely. In some cases, they may act by lowering the bacterial load or by shifting the relative abundance of commensal flora in the animals. In other cases, they appear to act by suppressing disease. In each of these applications, the drugs appear to apply some selective pressure.

Some of the agents used for growth promotion are not members of groups widely used in human medicine, but many are. For example, growth promoters have included β-lactams, tetracyclines, sulfonamides, macrolides, and fluoroquinolones.[195] In principle, use of these agents as growth promoters can lead to the emergence of bacterial resistance in food animals. In 2004, it was argued that the immediate benefits of antibiotic consumption by food animals (faster growth, healthier animals) needs to be considered more carefully before implementation of a "precautionary principle that is a nonscientific approach."[195] However, Denmark had banned the use of growth promoters in 1998–99, which enabled us to examine the effects of a ban on a country that practices extensive food animal production. In general, Danish farmers adjusted well to the ban, and the overall consumption of antibiotics by livestock and poultry dropped by 50%.[197] In 2006, the European Union withdrew approval for use of antibiotics as growth promoters.[196]

Antibiotic Contamination of the Environment Is a Byproduct of Usage

Many antibiotics are not easily destroyed when they escape into the environment. Consequently, we could be creating a huge reservoir of antibiotic-resistant microbes and resistance genes in our soil, rivers, and lakes by dumping medical and agricultural waste[198] (see Box 10-4). Although gene exchange among microbes occurs only rarely, even rare events can be significant when astronomically large numbers of microbes are involved.

Box 10-4: Contamination of Drinking Water

In March 2008, the Associated Press reported that a vast array of pharmaceuticals—antibiotics, anticonvulsants, mood stabilizers, and sex hormones—were found in drinking water supplies for more than 41 million Americans. In Philadelphia, officials identified 56 pharmaceuticals or byproducts in treated drinking water. These compounds included medicines for pain, infection, high cholesterol, asthma, epilepsy, mental illness, and heart problems; 63 pharmaceuticals or byproducts were found in the city's watersheds.

In September 2008, the Associated Press reported that the State of California warned its residents not to flush pharmaceuticals down the toilet or drain, because they may end up in drinking water. State and local officials joined with the U.S. Environmental Protection Agency for a "No Drugs Down the Drain Week." The program recommended that drugs be dropped at special collection sites or placed in the trash. Because procedures for removal of pathogens from the water supply do not completely remove antibiotics, we may need to add another layer to our water purification systems.

Hospitals are among the major polluters.[199] For example, in Hanoi, ciprofloxacin was present in untreated hospital waste-water at concentrations similar to *E. coli* MIC,[200] and a wide variety of substances has been associated with waste discharge in Taiwan.[199] Although several studies show that 80% to 85% of fluoroquinolone contamination is removed by waste treatment plants[200] that still permits large quantities to escape due to heavy consumption of these drugs. Moreover, much of the world lacks waste-water treatment. These considerations, coupled with the observation that some compounds, such as macrolides, are not readily removed by treatment plants,[201] leads us to conclude that hospitals are creating an environment that fosters emergence of antibiotic resistance.

Another source of environmental antibiotic pollution is likely to be the food animal industry. Whenever animals or fish are cultivated in dense populations, they become more susceptible to infection. Thus, the farmer and aquaculturist assert that they must use antibiotics to bring their animals to harvest. How much of this antibiotic use enters the environment and how long it persists is not known.

All serious gardeners know that rusts and molds can wreak havoc with ornamental and crop plants. Antifungal compounds counter such problems in many environments. These anti-mold agents are closely related to those used to treat life-threatening systemic infections in cancer and transplant patients. Cross resistance is commonly observed, and severely ill patients are now diagnosed with multi-azole resistant *Aspergillus fumigatus*,[202] a deadly mold transmitted by spores in the environment. In Holland, drug-resistant molds appear to be a by-product of agricultural azole use.[203] Azoles are also used to treat lumber—the drugs prevent discoloration of wood due to mold colonization. Such human activities may eliminate an entire class of antifungal agent from medical practice.

Clinical Resistance and Resistant Mutants Are Not the Same

We now turn to situations when use is necessary. In previous chapters, we pointed out that the clinical and microbiological perspectives of resistance are quite different. Clinical resistance means that an antibiotic is unlikely to cure infection when used according to a protocol approved by regulatory agencies. In this context, resistance is usually considered to be absolute: A pathogen isolate is either resistant or it is not. In contrast, a resistant mutant is a cell or strain that grows in the presence of antibiotic concentrations that block growth of so-called wild-type cells or the susceptible parent of the mutant. This laboratory definition implies no relationship with clinical cure: Many different resistance mutations may arise, and they may or may not confer clinical resistance. Indeed, a mutant containing several resistance mutations may have its growth blocked by antibiotic concentrations used clinically. Such an isolate would be considered clinically susceptible, because the patient can be treated successfully. The distinction between *absolute clinical resistance*, which is used to guide treatment with predetermined doses, and *resistant mutants*, which may be present in a susceptible isolate, is central to the following discussion of the emergence of resistance.

Dosing to Eradicate Susceptible Cells May Not Halt Emergence of Resistance

The chance that a mutant will arise depends on the size of the pathogen population. Consequently, reducing the number of pathogen cells in an infection should suppress the emergence of new mutants. Reducing the pathogen burden should also enable host defenses to more effectively eliminate residual subpopulations of resistant mutants. These considerations led to the conventional dosing approach in which the goal is to kill the bulk, susceptible population. With respect to resistance, this strategy is expressed as "dead bugs don't mutate."[204]

The conventional approach is inadequate because resistance can emerge during eradication of susceptible cells. Drug concentrations lethal to susceptible cells may permit the amplification of mutants present before treatment.[105, 205, 206] Although this idea is straightforward and data supporting it have been obtained using an animal model of infection,[105] showing that it applies with humans is not, because resistance emerges rarely on an individual patient basis. For most antibiotic-pathogen combinations, it is not known how many thousands of patients must be examined to determine the frequency at which a given dose enables resistance to emerge. Consequently, many students are still taught that emergence of resistance is best restricted by adjusting doses to cure infection. Below we briefly describe experimental results that counter that idea, and then we develop an antimutant dosing strategy.

Studies with cultured bacterial cells[207] and animal infections[105] show that an antibiotic can kill susceptible cells quite well and also permit resistant mutants to amplify. In one of the examples, an *S. aureus* infection of rabbits was treated with fluoroquinolone at concentrations inside the mutant selection window (between MIC and MPC). After several days of treatment, the dominant, susceptible population dropped by orders of magnitude, while growth of resistant mutants became obvious. Within a few more days the mutants dominated the population.[105]

Resistance also developed during eradication of susceptible cells in a human trial. In this experiment, a set of tuberculosis patients was identified in which patient noses were colonized by *S. aureus.*[108] When the patients were treated with rifampicin for tuberculosis, colonizing *S. aureus* was exposed to the drug. After 4 weeks of therapy, *S. aureus* was eliminated from noses in 90% (53/58) of the patients, whereas resistance emerged in the remainder (5/58). Molecular (DNA) strain typing showed that each of the 5 resistant isolates differed, which argued against patient-to-patient transmission as a source of resistance. Moreover, isolates obtained before and after therapy from each of the five patients carrying resistant *S. aureus* had identical DNA patterns. These data support the interpretation that resistance was acquired, that is, the resistant mutants derived from the susceptible population present before treatment. We conclude that treatment strategies aimed only at eradicating susceptible cells, that is, only to cure disease, may be inadequate at blocking acquisition of resistance. That conclusion is also reached from consideration of the mutant selection window hypothesis: To achieve cure with many pathogen-antibiotic combinations, drug concentrations are placed inside the selection window for long periods of time. Such treatments are expected to enrich mutant subpopulations and enable them to expand.

The key idea is that low doses, even if they seem to cure infection most of the time, lead to the emergence of resistance. This point is emphasized by another human study in which upper-respiratory carriage of penicillin-resistant *S. pneumoniae* was examined in small children (3 to 6 years old).[208] More than 900 French parents were questioned about the antibiotic use of their children for a month prior to sampling. Then *S. pneumoniae* was obtained from the children by oropharynx swabs and tested for β-lactam (penicillin and cefotaxime) resistance. The occurrence of resistant bacterial samples showed a strong correlation with low doses of β-lactam and with long treatment times (more than 5 days). The next question is how high must doses be to restrict emergence of resistance?

Keeping Concentrations Above MPC Restricts Mutant Amplification

By definition, antibiotic concentrations above MIC block susceptible cell growth; by analogy, concentrations above MPC should block mutant subpopulation growth, because MPC is the MIC of the least susceptible mutant subpopulation.[111, 209] Experimentally, when antibiotic concentrations are kept above MPC throughout treatment, mutant subpopulation amplification is inhibited with cultured *S. pneumoniae* and fluoroquinolones,[115] with cultured *S. aureus* and several antibiotics,[114, 210] and with an *S. aureus* infection of rabbits treated with a fluoroquinolone.[105] These data support MPC serving as a drug concentration threshold for restricting emergence of resistance.[110]

Keeping antibiotic concentrations above MPC throughout therapy is suitable for bacteriostatic antibiotics. However, it is more stringent than necessary for antibiotics that *kill* cells having a resistance mutation, because these antibiotics reduce the size of the mutant subpopulation. A smaller subpopulation has a lower chance of acquiring the second resistance mutation needed for growth in the presence of antibiotic at the MPC. An experimental demonstration of this principle is seen during infection of rabbits with *S. aureus*. Fluoroquinolone concentrations restricted mutant amplification when above MPC for only 20% of the dosing interval.[105] At present, the effect of lethal activity on emergence of resistance must be determined empirically for each antibiotic-pathogen combination.

MPC thresholds can be used to compare compounds within the same class for their ability to restrict emergence of resistance.[211] Pharmacokinetic data for approved doses are available, and they can be combined with MIC and MPC to determine whether concentrations of a particular compound are usually above the window, inside the window near the top, or inside the window near the bottom. Agents that reach higher concentrations relative to their selection window should be less likely to enrich resistant mutants.

Combining MPC with PK/PD Targets

A key problem is to identify minimal doses that restrict the emergence of resistance in large populations of humans. We next describe an approach based on the PK/PD strategy that is used to determine doses that will usually cure infection (see Box 4-4). We expect that as resistance becomes more prevalent, the PK/PD approach will receive the clinical tests required to make it standard practice.

For a variety of doses, we can measure drug concentration in humans at various times after treatment and determine the area under the time-concentration curve over a given time, such as 24 hours (AUC_{24}). AUC_{24} provides an integrated measure of drug concentration that is related to dose. The effective drug concentration depends on the susceptibility of the pathogen (MIC), which in Chapter 4, "Dosing to Cure," we expressed as AUC_{24}/MIC. When considering resistance, we are interested in the susceptibility of the least susceptible mutant subpopulation (MPC), because we seek to block mutant growth. If we divide AUC_{24} from a given patient by MPC, determined with the pathogen isolate from that patient, we can estimate the exposure (AUC_{24}/MPC) experienced by the least susceptible mutant subpopulation for a given dose. To determine whether a particular dose is high enough to restrict mutant growth, we can determine the value of drug exposure (AUC_{24}/MPC) that blocks outgrowth of mutants using dynamic in vitro models and experimental animal infections. If that value of drug exposure, which is an antimutant pharmacodynamic target, is exceeded by a given dose in the patient, it is likely that resistance will not emerge.

To estimate how well a particular dose would work for a large human population, AUC_{24} would be determined for a large number of patients treated with the particular dose, and MPC would be measured for a large number of pathogen isolates. Then the two values would be combined mathematically[212] to determine the fraction of patients for which the particular dose would reach the target, that is, the bacterial drug exposure that would severely restrict emergence of resistance. If that fraction of patients is deemed too small, higher doses could be modeled, which would require additional AUC_{24} measurements with many patients. The logic used to identify doses likely to restrict the emergence of resistance is the same as used to identify doses that cure infection (refer to Box 4-4), except that MPC is used to restrict the emergence of resistance and MIC to cure infections (see Box 10-5).

Box 10-5: Use of MIC Rather Than MPC to Restrict Emergence of Resistance

Efforts have been made to use MIC rather than MPC as an index for determining resistance-restricting conditions.[88, 212] A value of AUC_{24}/MIC that blocks growth of resistant mutants can be obtained experimentally using dynamic, *in vitro* models and animal infections: Gradually increase dose until no mutant amplification is observed. For consideration of a single isolate (a single infection), MIC serves as a surrogate for MPC because for any given pathogen isolate MPC will be a particular multiple of MIC.[114, 213] If the relationship between MPC and MIC were the same for *every* isolate, some multiple of MIC could replace MPC. Then that multiple of MIC, determined from large numbers of isolates, could be combined mathematically with AUC_{24} from large numbers of patients to determine the fraction of patients for which emergence of resistance would be blocked. Unfortunately, the relationship between MIC and MPC varies widely from one isolate to another.[214, 215, 216] Indeed, variability in the ratio of MIC to MPC is expected with patient isolates because some harbor mutations that confer little increase in MIC but a large increase in MPC and *vice versa.*[217, 218, 219] Thus, MIC cannot be used directly as a surrogate for MPC in calculations with large numbers of isolates. This issue is important because large MIC databases are already available while MPC studies are still few.[112, 220]

Combination Therapy Restricts Emergence of Resistance

When antibiotic resistance mutations are so protective that safe doses cannot maintain drug concentrations above MPC with a single agent, combination therapy can close the mutant selection window. In this dosing strategy, two or more compounds are sought that act independently: A resistance mutation against one does not affect the other. (This is called the absence of cross resistance.) In addition, the compounds should not interfere with each other. When the concentrations of the two compounds both exceed their respective MIC, the pathogen needs to acquire two concurrent mutations for growth. Because obtaining two concurrent resistance mutations is expected to occur rarely with bacteria, drug combinations will restrict emergence of resistance. Studies with tuberculosis show this to be the case.[221]

Because antibiotic concentrations in patients fluctuate, an additional issue exists: The concentration of both agents should be above MIC at the same time. If one compound is above its MIC while the concentration of a second is below its MIC, the pathogen will experience the equivalent of monotherapy with respect to the first compound. In principle, the two compounds should have similar pharmacokinetics, rising above and falling below MIC at the same time. When they don't, resistance will emerge for the antibiotic with the longer half-life, as seen in a clinical trial of the new antituberculosis agent rifapentine (see Box 10-6). In practice, perfect pharmacokinetic overlap is difficult to achieve.

Box 10-6: Resistance and Pharmacokinetic Mismatch

In a small clinical study, about 60 HIV-positive patients with tuberculosis were treated with a standard four-drug therapy for 2 months. Then the patients were divided into two groups. One received an additional 4 months of therapy with rifampicin and isoniazid (twice weekly), whereas the other received isoniazid and rifapentine (once weekly). Rifapentine is a derivative of rifampicin that has a long half-life. As a result, rifapentine concentration was expected to be above MIC for several days each week when the concentration of isoniazid was below MIC. This would be the equivalent of monotherapy for rifapentine. Rifapentine-resistant tuberculosis emerged in 4 patients, whereas the group receiving rifampicin produced no patient with resistant disease. In the latter case, rifampicin concentration was expected never to be above its MIC alone.[222]

These data indicate that good pharmacokinetic overlap between two compounds in combination therapy is important for restricting the emergence of resistance. Introduction of moxifloxacin, another long half-life compound, into tuberculosis therapy is expected to provide another test of pharmacokinetic mismatch ideas if the drug is used in combination with short-half-life agents: Fluoroquinolone resistance should arise quickly in HIV-positive patients.

Consideration of Resistance During Drug Discovery

Finding new antibiotic classes is difficult. Consequently, most of the work over the last 30 years has been devoted to refining existing classes to make the compounds slightly more potent with a broader pathogen spectrum. As we move forward one step at a time, the pathogens also move forward, one mutation at a time. Moving one mutational step is easy for the pathogens, because their populations can be so large that cells with one resistance mutation are abundant. Consequently, our efforts have only been holding pathogens at bay; new antibiotics are neither expected nor designed to restrict the emergence of resistance.

As pointed out previously, conventional antibiotic discovery efforts focus on finding compounds with low MIC and low mammalian toxicity. No antibiotic has been developed with restriction of resistance as a criterion. When resistance has been recognized as an important factor in therapy, multiple antibiotics having different molecular targets have been used to address the problem. (Examples are the current multidrug regimens for tuberculosis and HIV disease.) In some cases, it has also been possible to find specific inhibitors of resistance proteins. (β-lactamase inhibitors enable β-lactams to be effective.) The selection window hypothesis now leads to an approach for finding compounds that will themselves restrict the emergence of resistance (see Box 10-7).

Box 10-7: Antimutant Strategies for Antibiotic Development

The mutant selection window hypothesis leads to a way to find compounds that are less likely to succumb to resistance. The goal is to design a compound for which the selection window is closed, that is, MIC = MPC. In such a situation, subpopulations of first-step mutants would need to acquire an additional mutation to grow in the presence of the drug at the MIC. One way to find such agents is to test candidate compounds using a panel of known mutants. For example, with new fluoroquinolone-like derivatives we can test many different gyrase mutants that have the same genetic background. For each compound, the mutant MIC would be compared with the wild-type MIC to find compounds for which the two values of MIC are equal. Desirable agents should also suppress the amplification of spontaneous mutant subpopulations.[81]

Another strategy, which will work with some fluoroquinolones, is to identify agents that are dual targeting, that is, they have equal activity against the two targets of quinolones, gyrase and DNA topoisomerase IV. In this situation the MIC derived with the less susceptible target would equal the MPC. If that value also equals the MIC for the other target, MIC = MPC. Some fluoroquinolones come close to satisfying this criterion with *S. pneumoniae*, a major cause of pneumonia.[223, 224, 225]

Chemically joining two lethal compounds constitutes a third strategy. When derivatives of ciprofloxacin and rifampicin are joined, the resulting agent is active against both gyrase and RNA polymerase.[226, 227] Because the two activities are part of the same molecule, they are expected to have the same pharmacokinetics, thereby solving one of the problems associated with combination therapy (lack of pharmacokinetic overlap).

Perspective

A central reason for antibiotic overuse with human patients is the perception that the cost to the individual is small. However, the cumulative cost to the community is quite large. Philosophers call this situation the "tragedy of the commons." Curbing the urge to use antibiotics is difficult, because even doctors have that impulse. Physicians fear that they will fail to treat a treatable infection. However, cases are beginning to appear in which patient outcome is improved by using fewer rather than more antibiotics, and it is clear that commensal populations become resistant.[147] Such information enables doctors to conserve antibiotics while still meeting their primary obligation to the individual patient. Black market pharmacies, production of low-potency low-cost compounds, and self-medication still loom as major obstacles to lowering antibiotic use by consumers. Financial considerations drive agricultural use. We feed low doses of antibiotics to animals as "growth promoters," we treat overcrowded animals to bring them to market, and we spray antibiotics on crops to control bacterial and fungal pathogens. Nevertheless, an awareness of antibiotic resistance problems is developing. An example of this awareness surfaced in mid-2008. A major U.S. chicken producer thought that it was important to say that its chickens had not been treated with antibiotics. Thus, the chicken producer injected eggs with antibiotics instead of treating the young chickens.[228] At a different level are ecological effects from agricultural and hospital waste water. They are less clearly defined, but the principle is clear: Antibiotic exposure inside the selection window, regardless of the source, leads to antibiotic-resistant bacteria. Some of those bacteria are human pathogens, whereas others serve as starting points for movement of resistance genes into pathogens, as pointed out in Chapter 6, "Movement of Resistance Genes Among Pathogens." We can reduce each type of antibiotic use.

When antibiotic use is required, we argue that the conservative approach is to adjust doses to directly attack resistant mutants. The dosing philosophy of "cure-without-harm" needs to be shifted to "suppress-mutants-without-harm." A guiding theory is the mutant selection window hypothesis. Traditional dosing practices, which tend to place antibiotic concentrations inside the selection window for much of the dosing interval, can be corrected to keep concentrations above the window. Often, that requires doses that are higher than needed to cure disease. If those higher drug levels create side effects more often, patients are unlikely to accept increased personal risk to assure antibiotic availability for the population as a whole. Physicians are caught in the middle. On one hand, they are the stewards of antibiotics. At the same time, they are bound to provide the best possible treatment for their individual patients, which includes avoiding side effects. It is likely that implementing the use of higher concentrations to slow the emergence of resistance will require regulatory changes and additional safety tests at the higher doses. Unfortunately, continuing loss of susceptibility among pathogen populations will make the control of resistance increasingly difficult with each passing year. In the next chapter, we discuss the special case of influenza. This virus is different each year, and antibiotic resistance may disappear with the virus strain.

Chapter 11

Influenza and Antibiotic Resistance

Summary: Influenza is a viral disease that displays rapid dissemination of antibiotic resistance. The seasonal form of influenza moves around the world each winter, causing a mild, self-limiting infection for most persons. Vaccines are generally quite effective, but some patients fail to mount an adequate immune response. For such persons, antibiotics specific to flu virus have been developed. The adamantanes block productive infection of cells, whereas the neuraminidase inhibitors interfere with virus release and cell-to-cell spread. Resistance to the adamantanes has been extensive since 2005; resistance became significant for the neuraminidase inhibitors in the 2007–08 season. Pandemic influenza, exemplified by the 1918–19 outbreak and more recently by the 2009 H1N1 swine flu, sweeps around the world in a less predictable manner. Because vaccine can be produced only after human transmission is firmly established and the circulating virus strain is identified, antiviral agents are needed until a vaccine takes effect. For the 2009 H1N1 pandemic, vaccine was prepared and deployed; however, a lag of six months occurred. Thus, we know how long antibiotics will be needed. Avian flu, which has spread globally among birds, is monitored as a potential source of a new influenza pandemic. Adamantane resistance is widespread among avian flu isolates, and cases of resistance to neuraminidase inhibitors have been reported. The utility of these agents, which have been stockpiled, is unknown because resistance can emerge and spread rapidly with influenza.

In earlier chapters, we focused on diseases that tend to be problems in specific geographic regions. We now turn to a viral disease, influenza, which spreads so rapidly that it can affect the lives of almost everyone on the planet. Here, we briefly describe influenza biology and discuss how resistance may neutralize a major portion of our preparation for pandemic flu. Three general situations are considered: seasonal flu, human flu pandemics, and avian flu.

Seasonal Influenza Virus Is Controlled by Vaccines

Seasonal influenza, which moves around the world every year, is blocked by vaccines. Seasonal flu is thought to begin in wild waterfowl in Asia. From there, it spreads through domestic poultry and eventually reaches pigs. Viral reassortment occurring in pig tissues infected with both bird flu virus and human flu virus can produce a new version of influenza that is highly infectious to humans. (The viral genetic material comes as eight separate segments; when a cell is infected with two virus variants, progeny viruses acquire their eight genetic segments as a mixture, some derived from one parental virus and some from the other.) After the virus starts to spread among humans, it is isolated and used to prepare vaccine that is administered in the fall. Because the majority of infections occur months later in late winter (February and March in the Northern Hemisphere), there is time to prepare and distribute the vaccine. Unfortunately, new vaccines need to be prepared each year because the major circulating virus is different each year.

Antiviral Resistance Has Emerged Among Seasonal Influenza Virus

Flu vaccines are only 70–90% effective.[229] Consequently, antiviral drugs have been developed for persons unable to develop a good immune response to the vaccine. Influenza viruses fall largely into two general types called A and B. (In the 2008–09 season, 77% were type A, 23% were type B.[230]) Two antibiotic classes have been used for influenza A. The adamantanes (amantadine, rimantadine) inhibit influenza virus membrane protein-2 (M2). This protein is part of a channel required for passage of protons that help uncoat the virus. The adamantanes prevent flu virus from taking over the host cell. These drugs, which have been available since 1964, are ineffective with influenza B because this virus lacks the M2 protein.[230] The other antibiotic type, the neuraminidase inhibitors (zanamivir, oseltamivir [Tamiflu]), blocks the activity of the viral enzyme that breaks down glycolipids and glycoproteins. These agents prevent release of progeny virus from infected human cells. Neuraminidase inhibitors were designed for influenza A; influenza B tends to exhibit lower susceptibility. Zanamivir and oseltamivir target different sites on neuraminidase; consequently, the two drugs show little cross-resistance.[231] Oseltamivir is taken orally and has been applied throughout the world; zanamivir has been used largely in Japan.

In principle, antibiotic resistance emerges as a direct result of selection pressure acting on spontaneous mutants. Resistance can then be maintained in the virus population either by continued antibiotic pressure due to treatment or by the resistance mutation being located near other viral mutations that confer an advantage to the virus.[232] (Two mutations that are close to each other in a viral RNA segment tend to re-assort together and maintain a tight association.) In the latter case, resistant virus can spread to many persons who never receive antibiotic treatment.

In 2002–03, the circulating strains of seasonal influenza began to display adamantane resistance in Asia, perhaps stimulated by treatment of domestic birds.[230, 232] The emergence of resistance was exacerbated when the agents were added to over-the-counter cold remedies in Russia and China.[232] By 2005–06 amantidine resistance became extensive in the United States,[232] and by 2008 the viruses displayed such a high prevalence of resistance that the compounds were no longer recommended for treatment.[232]

Resistance to the neuraminidase inhibitors has also been appearing, even though these drugs have not been used extensively for domestic birds.[230] With clinical trials of oseltamivir, resistance was reported in 2% of treated patients (18% in treated children).[233] Such high numbers indicated that resistance would arise readily if the drug were widely used. However, prior to 2007 oseltamivir in untreated patients was rare (less than 0.3%).

In the 2007–08 season, the worldwide prevalence of oseltamivir resistance rose to 15% among seasonal influenza A H1N1 isolates.[231] (Subtypes are defined by differences in two viral proteins, hemagglutinin and neuraminidase, that are abbreviated by H and N, respectively.) Resistance was particularly evident in Norway (see Box 11-1). Later in 2008, resistant virus reached South Africa: Of 23 samples for which neuraminidase activity was tested, all exhibited oseltamivir resistance. Another 45 South African isolates were tested for a viral nucleotide sequence change associated with resistance. All tested positive for resistance. Because none of the South African patients had been treated with drug, spread of resistant virus appeared to be occurring. By late 2008, resistance was prevalent in the United States.[230]

In 2008-09, the common subtypes of seasonal influenza were H1N1, H1N2, and H3N2. These subtypes usually caused mild disease; consequently, antibiotic resistance was an issue mainly for immunocompromised persons and the elderly. However, understanding the emergence and transmission of resistance is important for managing more aggressive forms of the virus. One of the key points is that oseltamivir resistance in the seasonal H1N1 virus strain did not interfere with transmissibility nor did resistance correlate with treatment: Resistant mutants moved easily from person to person. Because new strains of seasonal flu appear each year, oseltamivir resistance may disappear when the current strain is replaced. However, it could reappear with little warning.

Box 11-1: Oseltamivir Resistance in Seasonal Influenza

Several influenza virus strains circulate at a given time, and often the dominant strain differs from country to country. In the 2007–08 season, an H1N1 strain was dominant in Norway, a country that used very little oseltamivir. When virus samples were analyzed, more than 67% (183/272) exhibited resistance. This resistant virus appeared to have spread in the absence of drug use.[234, 235]

continues

The H1N1 strain was still prominent the following year in the United States.[231] Surveillance carried out by the Centers for Disease Control[230] revealed in December 2008, that 24/25 isolates of influenza A H1N1 were resistant to oseltamivir. Other circulating virus types were still susceptible. These data establish that resistance can emerge and spread. In 2007–08, oseltamivir and amantadine resistance were in separate virus isolates,[232] which would make combination therapy a viable option. By 2009, virus carrying resistance mutations to both drugs had been found.[236]

Pandemic Influenza Can Be a Killer

A pandemic is an outbreak that spreads to multiple countries. The Spanish Flu pandemic of 1918–19 killed at least 40 million people worldwide within a 12-month period; more than 500,000 deaths occurred in the United States. An initial wave in the spring and early summer of 1918 was mild, whereas the second, in the fall of 1918, was severe. A third, less severe wave occurred in the winter and spring of 1919.[229] Flu victims were not limited to the old and infirm: Members of the healthiest age group, 15- to 34-year-olds, died in staggering numbers. Indeed, killing young adults seemed to be characteristic of the 1918–19 flu.[237] Smaller flu pandemics have also occurred. The 1957–58 Asian Flu killed nearly 1.5 million people worldwide, and the 1968–69 Hong Kong Flu had nearly 1 million victims. In 2009, a pandemic H1N1 strain originated from pigs in Mexico. To avoid embarrassment to Mexico, the pandemic was commonly called H1N1 Flu, which was confusing because much of the seasonal flu was caused by another H1N1 influenza virus. (The term Swine Flu lost favor because it led to erroneous notions about how influenza spreads.)

By mid-April, the 2009 pandemic had spread from Mexico to the United States and several other countries. Widespread panic did not develop. However, a lack of understanding surfaced in a variety of ways. Many people donned masks but failed to be properly fitted. That left large gaps between mask and face. Some countries banned pork imports even though the virus does not spread with food. Egypt experienced widespread slaughter of pigs for a virus that was spread directly from one human to another.

Experiences with influenza tell us what to expect from the next pandemic. Most important is the 6-month lag between virus emergence and vaccine availability. Second, obtaining and distributing sufficient vaccine may be difficult. We must not be lulled into complacency by the mildness of the 2009 pandemic. The panic associated with a highly lethal pandemic flu could overrun healthcare systems, cripple economies, and tear social fabrics as people fight for scarce medical resources. The Trust for America's Health, a nonprofit, nonpartisan foundation, estimated in 2008 that a pandemic comparable to the one of 1918–19 would result in 90 million illnesses and more than 2.2 million deaths in the United States alone (www.healthyamericans.org). The number of deaths globally would be 50 to 80 million.[238]

Two other recent disease outbreaks sensitized governments to the potential devastation of an influenza pandemic. In October 2001, the United States experienced a deliberate release of the anthrax bacillus, and about 1 year later an epidemic of severe acute respiratory syndrome (SARS) jolted China and Canada (see Box 7-6). These two small disease events extracted such a huge social, economic, and political toll that health officials designated pandemic illness a threat to national security.

In the United States, local and federal plans were developed to minimize the impact of large disease outbreaks. Those plans included stockpiling antibiotics. Until late 2008, little attention was placed on how antibiotic resistance could magnify the problem, and in the 2009 N1H1 flu pandemic, governments initiated limited distribution of oseltamivir. Fortunately, resistance with H1N1 pandemic influenza appears to have remained low (as of December 2009, the prevalence was about 1%).

Avian Flu H5N1 Is a Candidate for Deadly Pandemic Flu

Since 2003, avian flu has received considerable attention as a potential pandemic virus. This version of influenza A, caused by subtype H5N1, is endemic to Southeast Asia. In the late 1990s, it began its worldwide spread via birds, and between 2003 and September 2008 it caused 387 documented human deaths.[232] The human death toll was small, largely because transmission of virus occurred from bird to human rather than from human to human. The unsettling number was the crude mortality rate—a staggering 63%.[232] Flu experts expect the mortality rate to drop considerably when the avian flu virus adapts to human-to-human transmission, but how much is unknown. In 2008, the H5N1 virus was still spreading effectively among birds, and infection of humans required exposure to high inocula.

As pointed out, influenza viruses evolve rapidly; strains that are seemingly fit can mysteriously disappear. In the case of H5N1, the virus has persisted in the bird population since 1997, a long time for an influenza strain.[239, 240] During that time, many changes have occurred in virus recovered from poultry in Southeast Asia. For example, subtype H5N1 clade 2.3.4 became dominant in China in 2005 and then spread to northern Vietnam. (A clade is a group of viruses or organisms that descended from a single common ancestor.) In 2007, clade 2.3.4 replaced another Chinese clade that had successfully emerged in 2003.[241] As expected, the H5N1 virus is changing constantly, just like other versions of influenza virus.

Antibiotics May Play an Important Role in Pandemic Influenza

Strategies to counteract pandemic influenza depend largely on vaccine deployment, just as they do with seasonal flu. As of late 2009, no accepted vaccine for influenza H5N1 was widely available. (A vaccine had been made, but it was in short supply and was untested with human populations.) Vaccine makers were unsure which viral substrain to use for production. As pointed out, 6 months are required to prepare, test, and distribute a vaccine;[238] consequently, our control over pandemic influenza has a serious time lag. Quarantine and isolation can help cover the vulnerable time, but disruption to society could be enormous if even a small fraction of people stayed home from their jobs. It is likely that anti-influenza drugs will be needed.

To implement an antiviral plan, the Strategic National Stockpile began accumulating millions of antiviral drug doses (www.bt.cdc.gov/Stockpile/). Millions more were to be acquired by businesses, hospitals, and other essential components of society. The plans call for a phased prioritization of antiviral prophylaxis, which means that the first persons treated would be those who render critical services. The same approach applies to vaccine deployment.

Both the adamantanes and the neuraminidase inhibitors limit infection by susceptible influenza viruses if used early in infection. Late in infection, when the body reacts to the virus with flu-like symptoms (respiratory discomfort, fever, aches, and pains), the viral load is already quite high. At that point, reducing it with drugs is presumed to have little impact. Indeed, with highly lethal viral strains, the uncontrollable health problems come from a massive inflammatory response by the human body. Consequently, waiting for categorical symptoms before administering drug is considered an ineffective strategy. Therefore, antiviral *prophylaxis* is a key public health strategy for a pandemic of lethal influenza.

Antibiotic Resistance Occurs with Avian Flu H5N1

Strategic use of antiviral agents requires that a pandemic virus be susceptible to existing antiviral drugs. Susceptibility may be true at the beginning of an outbreak, but it cannot be known in advance because the pandemic virus is unknown. The government plan also assumes that the virus will remain susceptible to the drugs during widespread use. Given what we know about influenza viruses, this is problematic. Adamantane resistance has been developing for several years (see Box 11-2), as these drugs have been used prophylactically with poultry in China.[232] By 2007, 30% of the avian flu isolates were resistant to the adamandanes,[232] making the agents unlikely to be useful in a pandemic. In 2009, the neuraminidase inhibitors were the only option if avian flu had started to spread among human populations.[231]

Box 11-2: Spread of Adamantane-Resistant Avian Flu Virus H5N1

Adamantane resistance is associated with a serine-31 to asparagine change in the M2 viral protein. From 2001–2003, this change arose independently several times, and the resistant virus circulated in China. In 2003, resistant virus arose in a Vietnamese linage, and by 2004, it had spread to Thailand, Malaysia, and Belgium. Another independent strain arose in Indonesia in 2005. The latter strain spread to other regions in Indonesia and Sumatra. In 4 years, adamantane resistance emerged and spread throughout Southeast Asia.[232]

Neuraminidase inhibitors have generally been restricted to human use, and resistance mutations do not seem to drive the evolution of the virus.[232] (With seasonal influenza, the resistance mutation appears to have been a passenger on a viral genetic fragment that may have carried one or more other genes that facilitated transmission.) Thus, oseltamivir remains a good choice to stockpile. However, treatment is not completely suppressive for H5N1 virus with children or with immunocompromised persons.[232] In one study, oseltamivir resistance readily emerged; two of eight patients exhibited resistance during treatment.[242] These limited data indicate a high propensity for emergence of resistance, because with bacterial pathogens emerging resistance is rarely seen when individual patients are monitored, even with diseases considered to have a resistance problem.

Neuraminidase-mediated resistance tends to emerge from point mutations.[232] With oseltamivir and avian influenza virus, resistance following treatment has been attributed to conversion of histidine-274 to tyrosine. (This amino acid substitution fails to affect zanamivir treatment.[231]) As the virus evolves in birds, antibiotic susceptibility changes in complex ways due to the evolution of the target protein and other interacting proteins. That complexity can cause virus isolates from different times and different geographic locations to differ in drug susceptibility, which makes antibiotic resistance in avian flu difficult to describe in a simple way.

If avian flu causes a human pandemic, the problem of antiviral resistance will be exacerbated by arming citizens with oseltamivir and asking them to prophylax: The public cannot be expected to always get the timing and dosing right. When resistant virus emerges from infected persons taking antibiotic, those strains are likely to spread, as we have seen with seasonal influenza virus. To use the agents effectively, we need rapid, accurate methods for determining virus susceptibility. Current methods are still cumbersome (see Box 11-3).

Box 11-3: Detecting Viral Antibiotic Resistance

Detecting antibiotic resistance by virus growth requires more steps than with bacteria. Three methods are commonly used, although none was available in 2009 for routine commercial work with patient samples.[230] In one method, human or animal cells, grown in culture, are infected with virus in the presence or absence of antibiotic, and virus number is measured. Often, the drug concentration that reduces virus yield by 50% is determined (IC_{50}). A second method involves the target protein of the neuraminidase inhibitors (neuraminidase). Viral RNA is obtained from a patient sample, converted to a DNA form, and used to express the protein in vitro. The protein is then added to assays for neuraminidase, as are various amounts of the inhibitors. If the inhibitors fail to block neuraminidase activity, the virus is judged resistant.

A third method relies on knowing the particular viral nucleotide sequence change that causes resistance. In this strategy, the viral nucleic acid is examined for that change. A procedure called pyrosequencing is often used to detect nucleotide sequence changes. In this method, viral nucleic acid is extracted from a patient sample and then amplified by PCR (see Box A-3 in Appendix A, "Molecules of Life") after a step to convert viral RNA to DNA. The product of PCR is hybridized to a sequencing primer, and a new complementary DNA strand is synthesized stepwise, one nucleotide at a time. For each nucleotide to be incorporated, all four complementary nucleotides are added individually to the reaction mixture to determine which is put into the new strand. When incorporation occurs, a chemical called pyrophosphate is released. That release triggers several additional reactions that conclude with a flash of light that is recorded with a camera. (Light is produced only when the correct nucleotide is added to the growing DNA chain.) The process is repeated for each nucleotide in the template strand to obtain the nucleotide sequence. That sequence is then examined for the nucleotide change associated with resistance.

Bacterial Pneumonia May Create Another Resistance Problem

Influenza virus is only the beginning of the pandemic resistance problem. A major cause of flu-associated death is the bacterial pneumonia that follows flu.[238] Indeed, in the 1918–19 pandemic, most deaths appear to have been due to follow-up bacterial pneumonia.[243, 244] Although pneumonia-producing bacteria can be controlled with existing antibacterial agents, drug delivery systems will be challenged during a flu pandemic because we now rely on just-in-time supply chains. Developed countries have stockpiles of some antibacterials, but the optimal compound varies from one pathogen to another. For example, ciprofloxacin, which is quite effective with anthrax, is not recommended for pneumonia because its use with streptococcal pneumonia so often leads to resistance.[76] An alternative will be needed.

β-lactams, such as penicillin, and the newer fluoroquinolones are still widely used for infection by *S. pneumoniae*; consequently, they are likely to be administered prophylactically for pneumonia. That may create a problem with staphylococcal pneumonia, because MRSA is already penicillin resistant. Moreover, many HA-MRSA isolates are also resistant to fluoroquinolones. Thus, prophylaxis with β-lactams or fluoroquinolones will favor growth of MRSA. Vancomycin, one of the few effective antistaphylococcal drugs, requires intravenous injection, which cannot be easily administered in a pandemic setting. Moreover, vancomycin resistance occurs.[68] We conclude that antibiotic options may be quite limited.

Perspective

The striking emergence of antibiotic resistance in seasonal influenza virus is explained by the principles described in previous chapters. Factors likely to contribute are huge viral populations, agricultural and over-the-counter use of antibiotics, and dosing to cure rather than to prevent the emergence of resistance. What was responsible for oseltamivir resistance in the 2008 seasonal virus is unknown. In the case of neuraminidase inhibitors, exceptional care must be taken because the mutations causing oseltamivir resistance appear to move from person to person along with another change in viral RNA that improves transmissibility. Thus, continual selective pressure is not needed to disseminate resistant virus.

Some experts argue that pandemics are no more deadly than ordinary seasonal flu, as has been observed with the recent 2009 H1N1 pandemic.[245] However, the human death rate is only part of the issue. The perceived rate, not the actual one, will drive infrastructure disruption. Although plans to deal with an influenza pandemic appear orderly and well conceived, they assume that citizens will act in an organized, cooperative manner. That was true in 2009. Unfortunately, human nature has a way of intervening in life-and-death matters. As a society, we are so accustomed to drug-based healthcare that a massive run on antiviral agents may occur during a pandemic with a deadly virus. For example, a sudden demand for ciprofloxacin occurred following the anthrax outbreak in 2001. Despite pleas from health officials, extensive hoarding of ciprofloxacin occurred, and the drug quickly disappeared from pharmacy shelves. Personal stockpiling of oseltamivir for influenza also occurs. In a survey comparing seasonal influenza cases and oseltamivir sales from 2002 to 2006, the two generally coincided in time. However, sales of oseltamivir rose uncharacteristically in the autumn of 2005, a time when both avian flu and oseltamivir received a burst of media attention.[246] Another factor fueling uncertainty during a pandemic will be criminal elements. They sense easy profits, and counterfeit oseltamivir is produced for sale on the Internet. Low-quality antibiotics are a general problem for the emergence of resistance, because effective concentrations are low. (Totally inactive agents are the same as no treatment.) In the case of oseltamivir, a simple color test has been developed to determine whether a sample contains oseltamivir.[247] We conclude that pandemic influenza associated with severe disease will create a complex, challenging healthcare problem. Nevertheless, individuals can take action to minimize problems. In the next chapter, we sketch some of those actions for a variety of infectious diseases and resistance issues.

Chapter 12

Avoiding Resistant Pathogens

Summary: Individuals can avoid resistant pathogens by avoiding situations in which they might catch a resistant form of disease from another person or animal. When in such situations, precautions are advised: mosquito netting for malaria, cooking for food-borne diseases, no sharing of personal items for MRSA, and barrier protection for sexually transmitted infections. Good hand hygiene is essential. The key is to be aware of risky situations. Another principle is to limit the emergence of resistant pathogen populations by minimizing antibiotic use. (Don't use an antibacterial agent for a disease likely to be caused by a virus.) Self-medication is unwise with antibiotics: Physicians have more knowledge of antibiotics than patients, they know what diseases are common in the community, and they have access to laboratory tests that can sharpen the treatment. Finally, avoid enriching mutant subpopulations by using low doses and antibiotics of questionable potency (black market agents and old, left-over antibiotics in the medicine cabinet).

In previous chapters, we discussed ways for society to control antibiotic resistance. Now we turn to the individual consumer. With few exceptions, contracting an infectious disease, whether resistant or susceptible, is a matter of probability. This chapter focuses on lowering the odds.

Consumer Perspective Differs from That of Public Health Official or Manufacturer

As consumers of antibiotics we each have two tasks: cure disease and keep the drugs effective for next time. Our physician's view is similar. When we seek advice from a physician, *our* health is the primary concern. Public health officials have a different task: They are responsible for the world at large, not for any particular individual. Pharmaceutical executives have a third job: Make effective, high-demand products that meet safety standards and out-sell competitor products. These different objectives cause the same research data to be interpreted differently. For example, surveillance data obtained over several years with *S. pneumoniae* revealed that the prevalence of resistance to a fluoroquinolone went from 0.1% to 0.5% to 0.9% over the course of a few years.[248,249] When considering this situation, the physician was likely to say, "No problem, the prevalence of resistance is only 1%. You probably have a susceptible infection; antibiotic X should be fine." The same numbers stun the public health official, "Resistance is growing exponentially! In a few years, the antibiotic will be useless if we don't do something." The pharmaceutical company has many options: raise the dose, determine the mechanism of resistance and neutralize it, develop a combination dosing strategy with an additional compound, or sell the rights to the compound to another company. In this particular case, the prevalence of resistance leveled off at about 1%, perhaps due to the streptococcal vaccine, raising doses, and shifting to more potent quinolones. However, if these events had not occurred, the physician would conclude, perhaps when resistance reached 20%, that the odds were no longer good enough to prescribe the quinolones for streptococcal pneumonia. The public health community would write reports saying that the increase was predictable, and the pharmaceutical executive would consider a change in business plan. What can consumers do to protect themselves?

Avoiding Airborne Infection Is Difficult

The best protection from antibiotic-resistant pathogens is to keep away from them. That is particularly difficult with airborne pathogens, which spread in microscopic water droplets arising from talking, sneezing, and coughing. (Some pathogens, such as *M. tuberculosis*, remain infectious when the droplets dry.) Enclosed environments are problematic when occupied by individuals with active respiratory disease. For example, tuberculosis, which generally requires prolonged contact for transmission, has been traced to patrons of a particular bar and persons working inside ships. From time to time, we read about searches for passengers of airline flights after one is later found to have active tuberculosis. Sometimes the problem leads to a manhunt (see Box 12-1).

Box 12-1: Speaker's XDR-TB: A Clash Between Public and Personal Health

According to newspaper accounts, Andrew Speaker, a 31-year-old Atlanta lawyer, was diagnosed in 2007 with tuberculosis. He exhibited radiographic (X-ray) evidence of pulmonary tuberculosis, and samples taken from him contained *M. tuberculosis*. Moreover, the bacterial strain was thought to be extensively drug-resistant (XDR). However, *M. tuberculosis* was not found in a microscopic examination of sputum (respiratory tract mucus). Moreover, Speaker appeared healthy, showing no other symptom of tuberculosis. His doctors felt that it was fine for him to travel to Greece with his fiancé to get married. They deemed him a low transmission risk. Speaker and his new bride traveled from Greece to Rome for their honeymoon. Officials from the Centers for Disease Control (CDC) learned of the situation and notified Speaker that he could not fly home on a commercial airline due to potential transmission of the bacterium. He would have to remain in Italy for months until his treatment was complete. Speaker, fearing for his health, wanted to return to the United States for the extensive treatment. Because the CDC had placed him on a no-fly list to the United States, Speaker booked a flight to Montreal via Prague, rented a car, crossed the border into the United States, and drove to New York City. There, he checked into a hospital and contacted the CDC.

During his journey, Speaker had been a hunted man, and his travails made headlines around the world. When notified that Speaker had returned to the United States, the CDC flew him back to Atlanta under quarantine. He was then moved to the National Jewish Health and Research Center in Denver, which specializes in treating tuberculosis. The furor took a bizarre twist when a new series of lab tests, conducted at both the CDC and the Denver hospital, showed that Mr. Speaker had the more treatable MDR form of tuberculosis rather than XDR-TB. The behavior of Speaker and the CDC left many health officials shaking their heads about the handling of this especially dangerous disease.[250]

Influenza is another major respiratory disease that we worry about, as discussed in Chapter 11, "Influenza and Antibiotic Resistance." Vaccination is the conventional route to protection. However, the next pandemic influenza may not leave enough time for vaccine preparation and distribution. One alternative is treatment with agents such as oseltimivir. If administered early in infection, this drug can significantly diminish disease. (Oseltamivir was approved by the FDA in 1999, and by 2008, it had been used by more than 42 million persons in more than 80 countries.) In 2008, resistance became widespread among the seasonal virus strains, and treatment lost much of its value in some countries. Consequently, persons considering the use of oseltimivir should check for local resistance prevalence. If it is low, treat early and don't miss a dose. Another alternative is quarantine, a proven strategy for pandemic influenza. During the 1918–19 flu pandemic, a naval station in the middle of San Francisco Bay escaped infection by using armed guards to prevent access. Armed guards also protected several towns in Iceland. Such measures are extreme, but they work. Partial quarantine often applies to school children. During the 2009 pandemic, schools were closed to disrupt virus spread. (Child-to-child and child-to-parent transmission of viruses is common.)

Because transmission of influenza virus occurs via small air-borne droplets, it seems reasonable that face masks would limit disease spread. Virus also spreads from respiratory secretions transferred to surfaces, including hands, and then to faces. At the very least, masks keep hands away from the face. Masks of various types are being evaluated (see Box 12-2). Frequent hand washing with soap and water or the use of hand sanitizers during active flu season is advised. Pediatricians whose offices lack a sick-child waiting room might consider setting one aside. Because deaths associated with influenza often appear to arise from subsequent bacterial pneumonia, vaccination for *S. pneumoniae* may be a useful precaution, particularly for persons older than 65.

Box 12-2: Influenza and Face Masks

Face masks have two purposes. If the wearer is already infected, a mask could limit transmission to other persons. If the wearer is uninfected, the mask could keep the virus out. Although viruses are too small for their passage to be blocked by masks, tiny water droplets carrying virus can be trapped by the mask fibers. Masks have been studied extensively with professional microbiologists; however, little is known about mask usage by the general public. In one study, three masks were compared:

- Face piece against particles-2 (FFP-2), which is the European equivalent of the N-95 mask used in the United States
- Surgical mask
- Homemade mask prepared from tea cloth

Small particles were produced by burning candles, and their penetration of the masks was measured during either a 15-minute or 180-minute period. The FFP-2 mask protected healthy volunteers 25 times better than surgical masks, which were twice as effective as homemade masks. In a second experiment, the masks were fitted to a test head that was programmed to emit particles. The homemade mask provided little protection for the environment. The surgical and FFP-2 masks were only slightly better.[251] Thus, masks are likely to be more effective at protecting the wearer than the environment. Regardless of the intent, to be useful the mask must fit tightly. That poses a problem for adults with facial hair and for children (adult sizes do not fit properly). Also, remember that masks must fit over both nose and mouth.

With SARS (refer to Box 7-6 in Chapter 7, "Transmission of Resistant Disease"), we learned that hospital staff must exercise extreme care. (In Canada, about 20% of the transmission was traced to healthcare workers in hospitals.) Even though gowns and face protection were used, the virus still spread. Removal of protective equipment was a problem—hands tended to be contaminated, and special care was needed to avoid spreading virus. A protocol was developed that worked, but flaws were revealed by subsequent testing with a bacterial virus that was not infectious to humans. Knowing how to remove protective equipment may be important during an outbreak of influenza, especially for those providing home care with makeshift protection (see Box 12-3).

Box 12-3: Virus Transfer from Protective Clothing

If a lethal influenza pandemic occurs, many persons may be caring for family members at home. Although we may not have all the equipment needed, the procedures used in hospitals should be applied as best they can to slow the spread of infection. Personal protective equipment is generally worn for only a short time. In contrast, pathogens, such as SARS coronavirus and influenza virus, remain viable for hours. Infection can be spread by air, surface-to-hand contact, and hand-to-hand contact. Consequently, how one removes contaminated masks, gowns, and latex gloves is important.

In 2008, a report[252] appeared in which inadequacies in the current protocol were pointed out. Although the revised version has yet to be tested, it is instructive to consider the principles. In general, it is wise to wear two pairs of disposable gloves and tuck sleeves inside the outer gloves. The outer gloves are removed first by carefully peeling them off without touching the gown sleeves. Next the face protector (goggles or face shield) is lifted off by the strap with clean hands. (The inner pair of gloves are still on.) The face protector is placed in a container that can be sealed and disinfected. (Avoid touching potentially contaminated areas of the face protector to keep the gloves clean.) The gown is then removed and discarded without touching the exposed side. Next the respirator (mask) is removed, again by the strap, and discarded in a closed container that can be sterilized. Finally, the remaining gloves are removed, and hands are washed.

At all times, good hand hygiene is important to keep from spreading the pathogen. Covered, foot-operated bins are needed for removal of equipment to sterilization areas.[253] Boiling water is usually good for home sterilization; hospitals use pressure cookers called autoclaves.

Some airborne diseases are carried by dust. For example, hantavirus pulmonary syndrome is a deadly disease thought to enter the air from rodent saliva, urine, and feces. In the United States, the disease was first noticed in 1993 in the Four Corners area where a particularly wet winter enabled expansion of the field mouse population. The mice invaded buildings, and persons exposed to dried urine and droppings were infected. The case fatality rate can be high (50%). By 2006, more than 30 hantaviruses had been identified, and the disease was seen in the Balkans, Central America, and South America.[254,255] In regions of China, the virus has made its way into laboratory rats and from there to research personnel.[256] Pest control technicians now avoid sweeping mouse droppings because that creates a fine, virus-containing dust. (They advise washing away droppings with diluted bleach solutions.) The same precautions apply to bat droppings.

Precautions Can Be Taken with MRSA

MRSA emerged in the 1960s as a cause of healthcare-associated infections (HA-MRSA). Thirty years later, a different form of MRSA appeared in the community (CA-MRSA), generally among children and young adults. The organism invades wounds and even hair follicles. There, it establishes abscesses, which are often treated successfully by surgical drainage. Skin and soft-tissue infections are by far the most common consequences of CA-MRSA. Severe cases, such as deep-tissue infections and pneumonia, require antibiotic treatment.

Because MRSA is spread by contact, avoiding the pathogen is largely a matter of hygiene, as discussed in previous chapters. (Another example is given in Box 12-4.) Antiseptics may be useful when applied to skin abrasions as prophylaxis. Tests with liquid cultures of MRSA indicate that a mixture of benzethonium chloride with essential oils is more bactericidal than neomycin/polymyxin B or polymyxin B/gramicidin combinations.[257] Keep bacterial numbers down by cleaning abscesses, prevent spread by covering sores, discard used bandages carefully, wash hands frequently, use alcohol-based hand lotions, and avoid sharing potentially contaminated items (towels, bedding, washcloths, bar soap, razors, clothing, and athletic equipment). Much of this advice is supported by studies of disease outbreaks in prisons. For example, frequent showering and hand washing correlate with reduced risk of infection.

Box 12-4: MRSA and a Beauty Salon

In late 2004, a beautician in Holland experienced recurring infection with MRSA that required surgical drainage. After antibiotic treatment, she was declared MRSA-free (December 2005), but 3 months later, she tested positive for colonization. An epidemiologic study was performed that included 45 persons she contacted between July 2005 and December 2006. Fifteen persons had skin infections, and 10 of these individuals were colonized with MRSA. Overall, 11 persons were MRSA-positive, each with the USA300 strain. Two salon customers had skin lesions caused by MRSA; one was hospitalized. Waxing to remove unwanted hair was suspected as a route of bacterial transmission, but screening of 19 regular customers, employees, and waxing implements was negative.[258] Thus, waxing may not contribute frequently to transmission of MRSA.

When a person is infected with MRSA, infections can recur, probably from bacterial colonization of the patient's body. (Persons with hospital infections are often colonized by MRSA before the infection.) Common sites of colonization are the nose, mouth, and other body openings. Decolonization regimens sometimes help stop outbreaks, both in hospitals and in community settings. In addition to frequent bathing, antibacterial agents, such as chlorhexidine, are added to soap. The antibiotic mupirocin has been effective at decolonizing healthcare workers; however, recolonization is common. Consequently, routine antibiotic treatment is not recommended due to the risk of creating resistant colonization. Avoiding contact with large animals may be prudent: Horses are sometimes colonized with MRSA. (Large-animal veterinarians have a higher frequency of MRSA colonization than the general public.)[259] Companion animals (dogs) are reported to carry the same MRSA strain as veterinary staff.[260] Thus, treating a companion animal with antibiotics could, in principle, lead to the emergence of antibiotic-resistant bacteria that transfer to handlers.

Certain persons should be particularly alert for MRSA infection. Among the more obvious are members of a family experiencing recurring infection. Persons engaging in skin-to-skin contact (athletes) or living in crowded conditions (prison inmates, child-care attendees, military recruits) also have elevated risk. So do healthy newborn infants when the mother has a history of *S. aureus* infection. Farmers and food handlers constitute a new category of at-risk persons, because MRSA has entered the food supply chain (see Box 12-5).

Several other groups found to be at risk are children, young adults, native Americans, African Americans, and Pacific Islanders. Persons who are overweight need to take special care if folds of skin create moist regions.

Box 12-5: MRSA in Pigs, Their Handlers, and Food

In Northern Europe, where high-density food animal farming occurs, pigs are emerging as a reservoir for a type of MRSA called CC398. (This strain of MRSA is also found in horses and dogs in Austria and Germany.) CC398 exhibits resistance to six or seven drugs, including tetracycline, a compound commonly used to promote growth of food animals. CC398 first surfaced in 2004 in France where four pigs and a healthy pig farmer were colonized. The strain subsequently appeared in Holland and Denmark, two countries that employ a "search and destroy" policy to control MRSA. (Thirty-nine percent of Dutch pigs are thought to carry MRSA.) CC398 is traced by DNA methods, which now reveal pig-to-human and human-to-human transfer. In a study of veterinarians, 4% (179 total tested) carried the bacterium. Because Denmark has a sizable pig population (25 million are slaughtered each year), pigs could serve as a reservoir for human infection.[261]

In a 2006 report, roughly 80 raw meat samples, obtained from retail markets, had been tested for MRSA. Small samples (8 grams, about one-third of an ounce) were negative for surface contamination when tested by touching the meat to agar that was then incubated to enable bacterial growth. However, when the meat was tested by a more sensitive method (placement of the meat in broth culture medium followed by incubation), MRSA grew from 40% of the pork samples and one-third of the beef samples. Previous work had recovered *S. aureus* from meat products in 23% and 65% of samples collected in Switzerland and Japan, respectively.[262] Studies are also beginning to find MRSA in chicken.[263] Persons handling large quantities of raw meat should be careful when hands or arms get cut or burned.

Sometimes prior antibiotic use is a risk factor for MRSA infection. MRSA outbreaks have been a serious problem in jails and prisons where large numbers of persons are living in close proximity. A recent (2004–2005) examination of skin and soft tissue infections was performed with detainees at Chicago's Cook County Jail.[264] The study revealed that β-lactam use within the previous year correlated with a higher frequency of MRSA infection than infection with methicillin-susceptible *S. aureus*. This result led to the conclusion that empiric use of β-lactams at the jail is inadvisable.

Sexually Transmitted Infections Require Renewed Attention

A time window existed during the 1960s and early 1970s when unprotected sex was relatively safe. HIV had not surfaced, birth control devices guarded from unwanted pregnancy, syphilis and gonorrhea were easily cured by antibiotic treatment, and genital herpes had not become widespread. Those days are gone. Barrier protection is available, but condoms don't cover enough body to protect from CA-MRSA. Skin-to-skin transmission is making this pathogen an increasingly serious problem in some sexually active communities.[265]

Gonorrhea continues to merit attention. The prevalence of drug-resistant *N. gonorrhoeae* in Japan is 100% for penicillins, 70% for fluoroquinolones, 60% for tetracyclines, and 80% for macrolides.[23] As in many other countries, the pathogen is also found in the pharynx (throat). About 60% of Japanese women suffering from gonococcal urethritis or cervicitis also have throat infections. This apparent expansion of infection site may reflect a change in social behavior. In Japan, commercial sex workers charge less for oral sex. Moreover, oral sex is considered safer than vaginal sex. However, throat infections often have few symptoms, and they may be more difficult to diagnose and treat.[266] From a public health perspective, oral sex is a high-risk behavior.

Neisseria gonorrhoeae is inherently very susceptible to many antibiotics, and a single dose has been deemed sufficient for cure. Nevertheless, *N. gonorrhoeae* has become resistant to one antibiotic after another.[22] We may soon have no antimicrobial treatment for gonorrhea.[22] Because only one dose was used, resistance cannot be blamed on patients who failed to adhere to the treatment program. One idea is that extensive, and perhaps careless, use of the antibiotics for other reasons has applied unintended antibiotic pressure on *N. gonorrhoeae*. Another speculation emerges from experience with chloroquine-resistant malaria. For many years, malaria treatment was judged by abatement of symptoms, rather than eradication of parasites. Relapse was not distinguished from reinfection, which was thought to be common. As a result, resistance may have emerged from cycles of treatment, pathogen outgrowth, and retreatment. Then resistant parasites spread. With gonorrhea, urine or swabs from infected body parts are tested to indicate elimination of the pathogen. That may not be good enough, especially if throat infection is also involved. We may have been misled into thinking that single-dose treatments are adequate. From the standpoint of controlling antibiotic resistance, recommendations that multiple doses be administered for gonorrhea merit serious consideration.[267,268,269]

Arthropod-Borne Infections Are on the Move

Avoiding insect and tick bites is the obvious solution for arthropod-borne diseases. Most industrialized countries control their mosquito populations, making it unnecessary to sleep under insect-repellant-treated mosquito netting. But in many parts of the world, mosquito netting is important. Travelers to some countries, even recent residents of these countries, are advised to take antimalarial drugs as prophylaxis (see Box 12-6). For many viral diseases, such as yellow fever, effective vaccines are available. That is not yet the case for West Nile Fever.

Box 12-6: Malaria and Short-Term Travelers

In Europe, about 8,000 cases of imported malaria are reported each year. (From 1992 to 2002, more than 17,000 cases were recorded in children.[270]) Most cases of malaria in travelers occur among persons visiting relatives in African villages, as shown in Table 12-1. (Partial immunity that is seen with residents of countries having high malaria burdens is *rapidly lost* when these persons move to industrialized countries.)

Table 12-1 Malaria Risk in Travelers Returning to Sweden, 1997–2003[271]

Location	Risk per 100,000 Travelers
Arab countries and Iran	1.8
Indian subcontinent	62
East Asia	5.4
West Africa	302
East Africa	240
Central Africa	357
Southern Africa	46
Central America and Caribbean	1.3
South America	7.2

Travelers to regions where malaria is endemic can guard against the disease in several ways.[172] One is to restrict travel to cities, which generally have lower mosquito densities. A second is to remain inside buildings during hours of darkness, the time when the anopheles mosquito bites. If this is not feasible, wear long clothing and spray with a repellent containing DEET. Third, sleep in rooms protected by mosquito netting or controlled air flow (air conditioning). Fourth, treat with prophylactic drugs. Some agents, such as chloroquine, mefloquine, and doxycycline, affect only the blood stage of the disease; they do not prevent infection of the liver. Other agents, such as atovaquone-proquanil, interfere with both phases of disease.[172]

With respect to treatment, a study from Norway described nine patients with severe symptoms who were treated with artesunate, a derivative of artemisinin. All recovered quickly. Unfortunately, artemisinin resistance is beginning to emerge at the border between Cambodia and Thailand where a combination treatment of artesunate and mefloquine is routinely used.[55] Moreover, surveys in Africa reveal that in some countries, such as Senegal, artemisinin-containing medications are substandard.[272]

In the 1950s and 1960s, the United States routinely sprayed residential and urban communities with DDT to suppress mosquito populations. The treatment was highly effective, making mosquito- and tick-borne diseases uncommon. In 1972, DDT was banned due to its profound impact on wildlife. (The American bald eagle almost became extinct; see Rachel Carson's classic book *Silent Spring.*) Vector-borne diseases then rose significantly, and illnesses such as Lyme disease and West Nile Fever spread across the country. Despite environmental reservations, spraying programs have resurfaced, although DDT was replaced by more environmentally friendly pesticides.

Head lice are insects that spread easily from person to person. Like infectious agents, they cross all ethnicities and socioeconomic classes. Head lice infest 6 to 12 million children per year in the United States and are rapidly developing resistance to medications containing the botanical insecticide pyrethrum. In a study of 2,800 British children, two-thirds of head lice infestations were resistant to pyrethroid-based lice products. Ironically, pyrethrum drugs, such as permethrin, kill insects in much the same way as DDT. Some biologists believe that high-level use of DDT many years ago may contribute to the current emergence of permethrin resistance.[273] Many of the principles developed for emergence of resistance in microbes also apply to treatment of lice with insecticides.

Contaminated Food Is Common

At least 200 diseases are spread through food: In 1996, more than 76,000,000 cases of food poisoning occurred in the United States. Of this number, 353,000 required hospitalization, and 5,000 ended in death. Similar results, adjusted for population size, were reported for England and Wales (see Table 12-2).

Table 12-2 Disease Risks from Food-Borne Disease in England and Wales, 1996–2000[274]

Pathogen	Cases[a]	Hospitalization[a]	Deaths[a]
Campylobacter sp	338,000	16,000	80
Clostridium perfringens	168,000	700	177
E. coli 0157:H7	1000	389	23
Listeria monocytogenes	220	220	78
Salmonella (nontyphoid)	73,000	2600	209
Salmonella (typhoid)	86	35	0
Type of Food	**Cases[a]**	**Disease Risk[b]**	**Deaths[a]**
Poultry	503,000	104	191
Eggs	104,000	44	46
Red meat	287,000	24	164
Seafood	117,000	41[c]	30
Milk	108,000	4	37
Vegetables/fruit	50,000	1	14
Complex foods[d]	453,000	Not determined	181

[a] Number per year.

[b] Cases per million servings.

[c] Risk for shellfish is 650 per million servings.

[d] Mixtures of ingredients, source of infection not identified.

The vast majority of food poisoning is caused by viruses for which no antibiotic treatment exists. Nevertheless, bacterial infections are significant. Three types of bacteria stand out: *Campylobacter jejuni*, several *Salmonella enterica* variants, and *E. coli* 0157:H7 (see Box 12-7). *Campylobacter* causes most of the bacterial food-poisoning cases, but *Salmonella* is responsible for more deaths. *E. coli*, a normal inhabitant of the human digestive tract, became deadly when it picked up a gene for a powerful toxin from *Shigella*. This *E. coli* variant (strain 0157:H7) is uncommon, but serious. (The United States has 75,000 annual cases with 60 deaths per year, as reported in 1999.[275])

These three bacterial species contaminate many types of meat and poultry. Fortunately, cooking kills them. However, undercooking occurs, and the bacteria occasionally make their way into uncooked produce. Restaurant menus in some states are beginning to warn patrons that eating undercooked eggs, poultry, meat, fish, and shellfish poses a risk of food-borne disease. Moreover, some bacteria make toxins that are not destroyed by cooking. Thus, food poisoning is common enough to take notice.

Box 12-7: Campylobacter, Salmonella, *and* E. coli *0157:H7*

Campylobacter are short, cork-screw-like bacteria (see Figure 12-1a) that are responsible for much of the food-borne illness in industrialized countries. The two most common problem species are *C. jejuni* and *C. coli*. (*C. fetus* ranks second with children.) The organisms invade epithelial cells in the digestive tract, which then leads to fever, abdominal cramps, and occasionally bloody diarrhea. Some strains of *C. jejuni* produce a cholera-like toxin. Most infections are self-limiting, but a few are serious. Poultry is a major reservoir for *Campylobacter*. (The bacterium is a commensal in chickens.) Fortunately, freezing kills the bacterium.

Figure 12-1a *Campylobacter jejuni*. Scanning electron microscopy (magnification 11,734 times) shows that *C. jejuni* has a spiral shape.

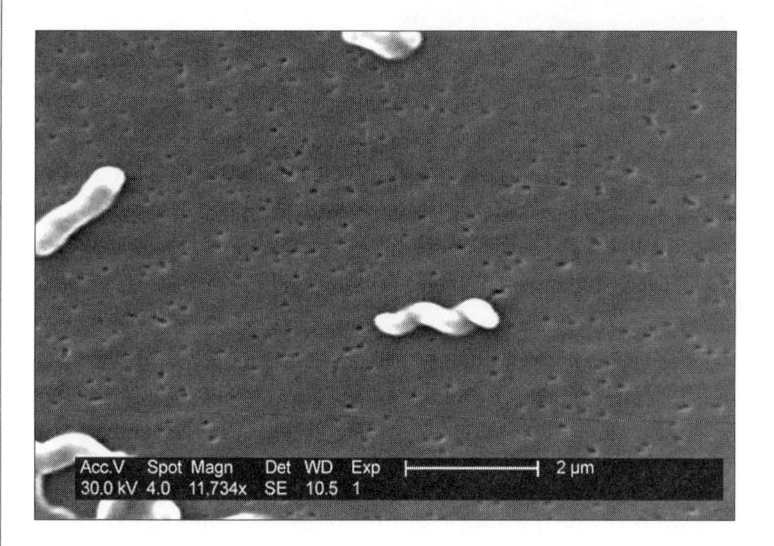

Public Health Image Library #5778; photo credit, Janice Carr.

continues

The genus *Salmonella* includes the bacterium responsible for typhoid fever and a large number of variants that cause nontyphoid food poisoning. These bacteria are small rod-shaped organisms (see Figure 12-1b). The nontyphoid types are named with three words because many serotypes exist that cause disease. (A serotype is a subspecies categorized by its capability to react with a particular immune serum; more than 200 serotypes have been identified for *S. enterica.*) Poultry serves as a major reservoir for nontyphoid *Salmonella*. The microbes spread around kitchens, often via sponges and cutting boards. Occasionally *Salmonella* enters a meat factory after the cooking stage, and it has even turned up in peanut products and cold cereal. In the United States, 1.4 million *Salmonella* infections occur each year. About 5% of these develop into bacteremia (rampant growth of bacteria in blood). In 2004, 15% were resistant to more than two antibacterial classes.[151]

Figure 12-1b *Salmonella enterica.* This scanning electron-micrograph shows a clump of bacterial cells obtained from a pure culture. Magnification is 8,000 times.

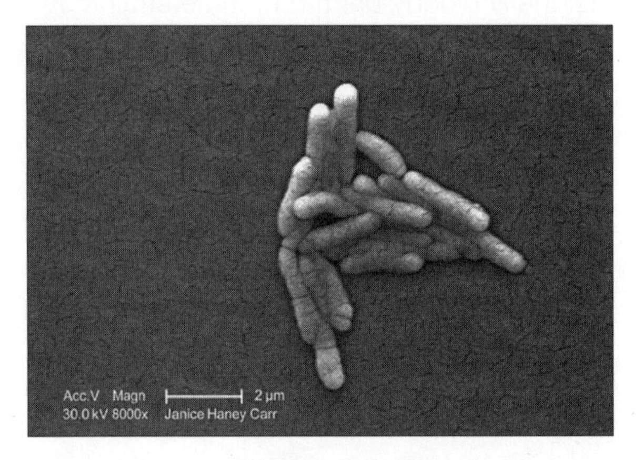

Public Health Image Library #10986; photo credit, Janice Haney Carr.

E. coli is another small, rod-shaped organism (see Figure 12-1c). It is a normal inhabitant of the mammalian digestive system. Bladder infections or cystitis occur when *E. coli* or other bacteria infect the urethra (urinary canal). Nearly 10 million persons are diagnosed with urinary tract infections each year, with women being much more likely than men to get such infections. (Nearly 50% of all women will experience at least one episode during a lifetime.) These infections, often caused by *E. coli*, are treated with antibiotics and are becoming increasingly resistant. In a study reported in 2001 involving women in

California, Michigan, and Minnesota, 22% of the *E. coli* strains were resistant to trimethoprim-sulfamethoxazole, a first-line antibiotic treatment for bladder infections. Alarmingly, nearly 50% of drug-resistant infections from these three distinct geographical regions were caused by a previously unrecognized strain of multidrug-resistant *E. coli*.[276] The 0157:H7 form of *E. coli* is a special case because it is a dangerous food pathogen. Ruminants, such as cattle, serve as a reservoir for *E. coli* 0157:H7, which enters beef products during processing. This organism has also been found in milk, produce (vegetables), and occasionally apple cider.

Figure 12-1c *Escherichia coli* O157:H7. This group of bacterial cells was obtained from a pure bacterial culture and examined by scanning electron microscopy. Magnification is 6,836 times.

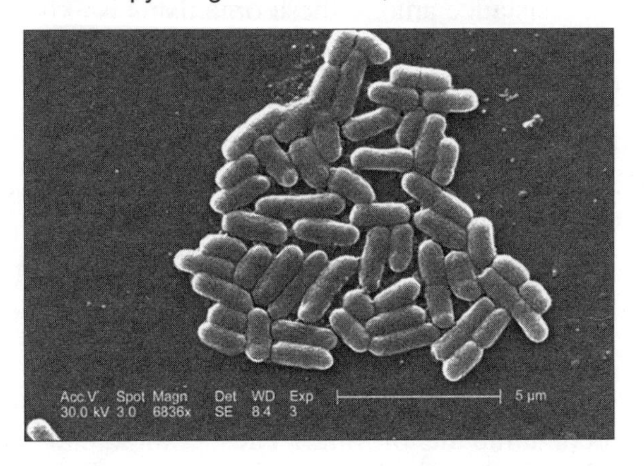

Public Health Image Library #8800; photo credit, Janice Carr.

E. coli is often considered to be an indicator of fecal contamination. (High levels of coliforms, bacteria that include *E. coli* and *K. pneumoniae*, cause health departments to close ocean beaches to protect swimmers from sewage.) In U.S. grocery stores, *E. coli* contaminates about half of the pork and 90% of the chicken. Most of that contamination probably comes from animals rather than humans, because human waste is carefully processed. (Infected food handlers are blamed for only 4% of the food poisoning cases.) The animal food industry is large. (In the United States, about 800,000,000 chickens are hatched per month for cooking, along with 340,000,000 egg layers; turkey hatchings account for another 270,000,000 birds.) All these animals defecate. Indeed, farm animals contribute much more fecal material than humans to the U.S. environment.

The massive amount of contamination makes it difficult for even very clean food processing plants to remove all bacteria. Consequently, it is not surprising to find meat products contaminated with bacterial pathogens, such as *Salmonella*, that flourish in these nutrient-rich environments. Nor is it surprising that some of those organisms are antibiotic resistant, because food animals are treated with antibiotics. Indeed, nucleotide sequence analysis of food contaminants now supports the argument that antibiotic resistance emerges in animals.[277]

Although issues of animal waste and resistant *E. coli* and *Salmonella* apply to vegetables irrigated with contaminated water, many other food pathogens derive directly from humans. Among these are *Shigella, Salmonella typhi,* and hepatitis virus type A, agents that often reach food during processing or preparation. *Listeria*, which is particularly problematic for pregnant women, enters the food supply at the processing step: It contaminates refrigerators in food processing plants. Antibiotic resistance among these organisms is expected to derive largely from medical practice. Thus, finding ways to reduce the prevalence of resistance depends on the particular pathogen.

Antibiotic treatment of food animals contributes to the emergence of resistant bacteria in humans. (Examples are described in Box 12-8.) Even as far back as 1998, about 30% of the *S. enterica* Typhimurium cases were multidrug resistant.[278] Occasionally, outbreaks are extensive. For example, two waves of *Salmonella* infection emerged in pasteurized milk from an Illinois dairy.[279] One involved almost 169,000 cases, the other 198,000. Overall, 12 outbreaks in pasteurized milk occurred between 1960 and 2000.[280] (Pasteurization is a heating process that kills most microbial contaminants of milk.) Thus, commercial dairies, which have an excellent product safety record, are subject to the same concerns as other parts of the agriculture industry.

Box 12-8: Animal-to-Human Transmission of Resistant Salmonella

Salmonella enterica serotype Typhimurium is the most common serotype isolated from humans. A survey (1997–2003) examined all isolates obtained from patients in the state of Minnesota for resistance and molecular typing. The same sampling method was also performed on clinically ill farm animals. Of the 1,028 human isolates, 44% were resistant to one or more antibiotic, and 29% were multidrug resistant. Many (271) bacterial subtypes were identified by DNA-based testing. The animal isolates, mainly from cattle and pigs, also represented many subtypes and exhibited a high prevalence of resistance to one (89%) and multiple (81%) antibiotics. Both human and animal multidrug-resistant isolates fell largely into the same two clonal groups, consistent with transmission of *Salmonella* from animals to humans.[281]

In a case study, a 12-year-old boy living on a farm in Nebraska contracted a *Salmonella* infection that was resistant to 13 different antibiotics.[282] One of the resistance markers (ceftriaxone) was very uncommon for *Salmonella* in the United States, which caused health authorities to pay attention to the case. Two weeks earlier, the boy's father had treated sick calves for severe diarrhea, and samples from the calves matched those in the boy by a variety of genetic tests. Although antibiotics were not required to treat the boy's infection, the investigators examined the boy's medical records. It turned out that he had just completed antibiotic treatment for another infection. The prior treatment may have predisposed him to infection by drug-resistant *Salmonella.* (Antibiotic treatment would have reduced the commensal bacteria in his digestive tract and favored growth of the drug-resistant strain of *Salmonella.*)

One of the negative consequences of antibiotic-resistant *Salmonella* is increased severity of human disease. Food-borne pathogens generally become resistant from use of antibiotics with food animals. Fluoroquinolones serve as an example. These drugs were introduced into veterinary medicine in Europe in the late 1980s and into the United States in 1995. Fluoroquinolone-resistant *Salmonella* infections soon began to surface. In Denmark, where all persons are tracked by a national registry, it is possible to examine the long-term consequences of being infected with a resistant strain. In one set of studies,[283] the mortality rate for Danish persons following infection with antibiotic-susceptible *Salmonella* was twice that of the general population. After infection with a strain resistant to five antibiotics other than fluoroquinolones, the mortality rate rose to five times that of the general population. Fluoroquinolone resistance alone correlated with a death rate ten times higher, and fluoroquinolone resistance plus resistance to five other antibiotics increased the rate to 13 times that of the general population. An important aspect of this work was correction of patient data for differences in other illnesses.

Three factors contribute to increased mortality from drug-resistant food infection. One is empiric treatment, which is started before resistance is known. Such treatment is likely to be ineffective when the bacteria are already resistant to the drug chosen for treatment. Consequently, effective treatment is delayed.

Another is the availability of fewer treatment options for drug-resistant infections. Third, secondary treatment options often have more severe side effects.

A separate issue is the chance of contracting a *Salmonella* infection after taking antibiotics for a different infection. Several studies examined sporadic (nonoutbreak) diarrhea caused by *Salmonella*.[284] A risk factor for infection with multidrug-resistant *Salmonella* is antibiotic treatment during the preceding 4 weeks. The risk is greatest if treatment is with a drug to which *Salmonella* is already resistant. Because some isolates carry eight different resistance genes, being treated with an ineffective antibiotic is likely. The phenomenon is explained as patients initially becoming infected with a small number of bacteria whose population enlarges. (Antibiotic treatment favors resistant bacteria and reduces commensal bacterial populations.) The result is antibiotic-resistant diarrhea. The average time between starting antibiotic therapy and onset of illness is about 2 weeks. However, outgrowth of resistant bacteria can occur sooner, because eating runny eggs or uncooked ground beef during the 5 days before onset of illness has been correlated with resistant disease.

The current strategy for protecting consumers is to keep the pathogens out of the market place. With cattle, the hide is a major source of contamination. Consequently, washing hides and carcasses with disinfectants is a common practice. As a further precaution, the meat can be heated or irradiated during processing. It can also be treated with a carbon monoxide-anaerobic gas mixture. This process greatly increases the shelf-life of beef by preventing putrefaction and contamination with *E. coli* 0157:H7. (Some consumers fear that this gas treatment will be used to deceive them about the age of meat products.)

Further down the supply line is control of flies (see Box 12-9). Temporal correlations between fly populations and *Campylobacter* outbreaks in Europe led to the hypothesis that flies from animal production areas carry the bacterium.[285] Placing screens on chicken houses lowered the fraction of flocks infected with *Campylobacter* from 51% to 15%.[286]

Box 12-9: E. coli *0157:H7 at County Fairs*

E. coli 0157:H7 is now widely distributed among livestock, particularly cattle (13% of feedlot cattle; at slaughter, the number rises to 28%). For pigs, the number is lower, about 2% at the time of slaughter. In addition to possible contamination of the food supply, animal handlers are exposed to the bacterium. Many more people are potentially exposed at state and local fairs where millions admire the animals. To assess risk, sampling for *E. coli* 0157:H7 was conducted at fairs in the Midwestern United States. Feces of 13% of the cattle contained the bacterium (1.2% with pigs). Because these animals are carefully tended, it is unlikely that the prevalence of *E. coli* 0157:H7 can be reduced by improved hygiene. The quality of hand hygiene by the handlers and visitors to fairs is likely to be important for reducing animal-to-human transmission. In the same study, flies were trapped and examined at 21 fairs. Roughly 19% of the flies carried *E. coli* 0157:H7. It is important to point out that no case of disease was specifically attributed to flies. But the study[287] does point to an area requiring caution.

Resistance can also be reduced by eliminating antibiotics from agricultural use. A 2005 legal ruling removed fluoroquinolones from veterinary use with poultry (refer to Box 8-4 in Chapter 8, "Surveillance"). This ruling was significant, because it allowed the Food and Drug Administration to consider resistance during the antibiotic approval process. Despite moves in Europe and the United States to eliminate fluoroquinolones from poultry, fluoroquinolone-resistant *Campylobacter* will probably persist, because most of the resistance mutations have little fitness cost. Moreover, countries in Southeast Asia continue to apply fluoroquinolones to poultry and other food animals. Travelers should beware. (Thirteen percent of the *Campylobacter* and *Salmonella* infections in the United States are in international travelers.) Extremely resistant *Salmonella* (see Box 12-10) is an example of an Asian issue.

Box 12-10: Extremely Resistant **Salmonella**

Since 2000, Taiwan has experienced a rapid increase in human infections with a serotype of *S. enterica* called Choleraesuis.[288] The bacterium is very invasive, often causing sepsis and even attacking the aorta. It is also resistant to several antibiotics, including ampicillin and fluoroquinolones. A quarter of the patients arrive at the hospital in shock with a disease that is difficult to treat. In Taiwan, over 70% of the persons infected with Choleraesuis have bacteremia. Most of these patients are also ill for other reasons (diabetes, HIV disease, immunosuppression) that may make them particularly susceptible to *Salmonella*. Health authorities attribute the infection to eating pork, because this form of *Salmonella* is the major type found in pigs. Moreover, in Taiwan fluoroquinolones are widely used on pigs.

Legislation concerning food contamination occasionally needs careful crafting to have an impact. In Denmark, the public was thought to be sensitive to contamination of meat by *Campylobacter*. However, legislation to stop contaminated chicken from reaching market initially had little effect on purchasing habits. *Campylobacter*-free chicken was not a big seller. To solve this problem, legislation was coupled with freshness. (Chicken could not be sold as fresh unless it was free of *Campylobacter*.)

Food poisoning is one of the easier resistance problems to solve. Cooking food thoroughly is the best way to reduce the bacterial burden. Also, good kitchen hygiene is important for avoiding the transmission of bacteria from raw meat to ready-to-eat food. That means washing hands and any implements that have touched uncooked meat. Avoid sponges. Wiping up raw meat drippings with a sponge can create a reservoir for future spread of bacteria. Use a paper towel and dispose of it properly. Keep flies away because they can mechanically transfer *Campylobacter*.[289] Prompt refrigeration of leftovers is also a good idea, because most bacteria grow more slowly at reduced temperature. Other problems arising from use of raw eggs can be old-fashioned eggnog, a favorite at Christmas time, and licking bowls and spoons after preparing cake batter.

Avoid Rounds of Treatment Interspersed with Pathogen Outgrowth

To this point in the chapter, we have focused on situations in which resistance arrives with the infecting pathogen, essentially from the outside. Emergence of resistance within a given person is a different issue. In some cases, such as treatment of syphilis with penicillin, resistance emerges rarely if at all, and one need dose only high enough to remove symptoms. In other situations, dosing strategies need to block mutant growth. For example, azithromycin was thought to be effective as a single-dose treatment with syphilis,[290] but resistance to azithromycin is now a growing problem.[291] With tuberculosis, half of the patients treated only with streptomycin or isoniazid develop drug-resistant infection. HIV-positive patients with tuberculosis develop rifapentine resistance, and those with MDR tuberculosis develop fluoroquinolone resistance. At the beginning of Chapter 5, "Emergence of Resistance," we discussed a case of oxacillin-resistant, vancomycin-nonsusceptible *S. aureus*, and in Chapter 10, "Restricting Antibiotic Use and Optimizing Dosing," we described patients who acquired rifampicin-resistant *S. aureus* when treated for tuberculosis. In the latter study, genetic analyses pointed strongly to susceptible bacterial populations becoming resistant in the patients during treatment. Other cases exist with fluoroquinolone-resistant pneumococci.[104] Thus, the principles of mutant amplification apply to a broad range of pathogen-antibiotic combinations.

A key is to avoid treatment interspersed with periods of pathogen population expansion. With each successive treatment episode, the resistant fraction of the pathogen population increases. Eventually, the disease may become untreatable by that antibiotic. (In the laboratory, gradual enrichment of mutant subpopulations is easily seen after repeated treatment, dilution, growth of survivors, and retreatment.) In humans, mutant enrichment is common in patients with uncontrolled HIV infection who are treated for recurring bacterial and fungal infections. Gradual mutant enrichment is also seen during treatment of *Pseudomonas aeruginosa* in the lungs of cystic fibrosis patients. These bacteria cannot be completely removed, and eventually resistance emerges to one drug after another. With other diseases, enrichment of resistant mutants over many years of recurring infection is not well documented, but we may not have looked hard enough for it.

Consume Only with Sound Indications, Choose Optimal Antibiotics

Proper antibiotic use means, in part, that antibiotics should be used only for clear medical indications. Antibacterials should not be used for viral infections, and antivirals should not be used for bacterial infections. Proper use also means no self-medication with prescription drugs. Self-medication suffers from the lack of access to laboratory tests, knowledge of surveillance studies, and inexperience at guessing probable causes of infection. Self-medication is often associated with low doses due to out-of-date antibiotics (reduced potency arises from oxidation of left-over prescriptions), to splitting pills (sometimes prescriptions for HIV infections are shared in resource-limited countries), and to reduced treatment time. Black-market agents may be of low quality, which increases the chance that resistance will emerge.

In some cases, as with treatments for vaginal yeast infection, patients often self-diagnose and self-treat with over-the-counter products. To restrict the emergence of resistance, such products must be used according to the manufacturer's recommendations. When used responsibly, these topical treatments are expected to generate high local concentrations, making it unlikely for resistance to be significant.[292]

Antibiotic choice may be important. For example, long half-life compounds may remain at mutant-selective drug concentrations for long periods of time. Cross-resistance is another choice problem. Two compounds can have different molecular targets and still show cross-resistance at the level of efflux (pumping drugs out of cells, see Table 12-3). Thus, low-dose treatment or use of marginally effective antibiotics that initiate emergence of resistance for one compound may also start the pathogen on the path to resistance for several other compounds. The conservative approach is to use the most potent member of an antibiotic class.

Table 12-3 Multidrug Resistant Efflux Systems[118]

Efflux System	Drugs Displaying Efflux-Mediated Resistance
AcrAB-TolC[a]	aminoglycosides macrolides lincosamides ketolides glycylcyclines fluoroquinolones oxazolidinones triclosan[c] quaternary ammonium compounds[c] phenolics[c]
MexAB-OprM[b]	tetracyclines glycylcyclines β-lactams aminoglycosides fluoroquinolones triclosan[c]

[a] Data are for *Escherichia coli.*

[b] Data are for *Pseudomonas aeruginosa.*

[c] Biocide.

Whether we should dose to prevent resistance may depend largely on the size of mutant subpopulations. If the size of the resistant mutant subpopulation is low, doses sufficient to remove only the susceptible cells may be adequate. If, however, resistant subpopulations are large, we must dose to block their outgrowth. Determining mutant subpopulation size is straightforward in research laboratories,[111,293] but not in clinical settings. Thus, in most cases, we will not know the mutant subpopulation size. The conservative approach is to dose to suppress resistant mutants, especially with anyone having a weakened immune system: the very young, the old, persons taking immunosuppressing drugs, and persons with diseases that lower immune function.

A new problem may be arising from the widespread use of household disinfectants. They are heavily advertised to have antibacterial activity. As they reduce the load of bacteria in our homes, they exert selective pressure that favors survival of organisms having mutations in genes involved in drug efflux. Although it is argued that disinfectants are used at such high concentrations that even resistant mutants cannot survive, we are aware of no study showing that routine household use of disinfectants prevents the emergence of resistance.

Hand sanitizers raise different issues. Soaps containing the biocide trichlosan are similar to plain soap at reducing levels of bacteria on hands.[294] Thus, no clear advantage is gained by "bactericidal" soaps. Nevertheless, during cold and flu season, alcohol-based sanitizers are likely to be useful for limiting the spread of viral disease, especially if they are used in addition, rather than instead of, soap and water. We reiterate that many hand sanitizers are ineffective against spores, such as those formed by *Clostridium difficle*. Thus, the reason for hand hygiene needs to be considered when deciding the appropriate procedure.

Perspective

Living organisms are complex, and ideas that may seem to be common sense do not always turn out to be useful. Consequently, physicians are taught to use evidence-based medicine, and many studies are performed to test what may seem to be obvious. For example, studies in prisons show that persons who shower more often are less likely to have problems with skin infections caused by MRSA. But surprises come along. For example, sharing bar soap is a risk factor for MRSA infection. Problems arise when decisions must be made in the absence of direct evidence. Many aspects of antibiotic resistance fall into this category, largely because diagnosis of infections at an early stage is difficult and because drug susceptibility testing is not routinely performed for certain types of organisms (for example, yeast and mold infections). Thus, we must sometimes rely on general principles. We can try to develop appropriate habits and attitudes, such as good hand hygiene and cooking potentially contaminated food. Microbial awareness in the kitchen can become second nature. We also need to consider the possibility that antibiotic treatments place us at risk for adverse effects (see Box 12-11). Toxic side-effects are well known (see Table 12-4). Less appreciated and less thoroughly documented are the consequences of current antibiotic consumption for future resistance problems (see Table 12-5). We expect the list in Table 12-5 to grow as our knowledge of resistance increases.

Box 12-11: Adverse Effects of Antibiotics

Data collected between 2004 and 2006 showed that more than 140,000 persons per year in the United States visited emergency rooms (ERs) due to adverse reactions from antibiotics, accounting for 20% of ER visits for all prescription drug-related side effects. Because only 16% of all prescriptions were for antibiotics, antibiotics are *not* safer than the average drug. Antibiotics differ from other drugs with respect to the type of adverse effect: Almost 80% of the antibiotic adverse effects are for allergic reactions, which cannot be reduced by limiting prescription errors. (Allergic reactions may not be easily predicted.) For other drugs, ER visits are usually for overdose or medication errors, not for allergic reactions. Data for several drugs are shown in Table 12-4.

Table 12-4 Toxic Side-Effects of Antibiotics[295]

Antibiotic	Cases of Side Effects/10,000 Patients
β-lactam	19
Fluoroquinolone	9
Sulfonamides, trimethoprim	19
Macrolides, ketolides	5
Tetracyclines	5
Vancomycin	22
Nonantibiotic	
Warfarin, insulin, digoxin	21
Anticoagulants, antiplatelet agents (aspirin)	3

A second type of adverse effect, which is unique to antibiotics, is increased risk for subsequent infection with antibiotic-resistant pathogens (see Table 12-5). One explanation for this phenomenon is that antibiotic treatment alters the microbial ecosystem and favors growth of antibiotic-resistant pathogens.

Table 12-5 Risk for Subsequent Resistance

Antibiotic	Consequences
Clindamycin, β-lactam (cephalosporin), or fluoroquinolone	*C. difficile* overgrows other digestive tract flora and causes a serious form of diarrhea.[142, 296]
Fluoroquinolone	Fluoroquinolone-resistant pneumonia.[297] Moreover, fluoroquinolone-resistant *E. coli* in the digestive tract could contribute to resistant urinary infections.

Many antibiotics	Increased risk for antibiotic-resistant *Salmonella*.[284]
β-lactam with young children	Increased carriage of penicillin-resistant *S. pneumoniae*.[208]
β-lactam	Increased risk of MRSA compared to MSSA.[264]
Azoles	Increased risk of azole-resistant fungal infections during immunocompromise (for example, anticancer therapy).

The concepts described in *Antibiotic Resistance* illustrate why many infectious disease experts consider resistance to be inevitable: Pathogen populations are large, mobile genetic elements introduce resistance at high frequency, our societies encourage massive use of antibiotics, and our treatment strategies seem almost designed to selectively enrich mutant subpopulations. Even hospital personnel are careless about hand hygiene.[179] The last three features we can correct; we are optimistic that changing awareness and human behavior will greatly lengthen antibiotic lifespan. In the following Afterword, "A Course of Action," we describe a plan.

Afterword

A Course of Action

The continuing hope of many infectious disease experts is that the pharmaceutical industry will temporarily stave off the inevitable by producing new classes of antibiotics. That would keep us one step ahead of the pathogens. However, three fundamental problems stand in the way. First, if new, highly effective compounds were produced, the medical community would restrict their use to avoid loss of efficacy through emergence of resistance. In the absence of sales, little profit can be made. Second, staying only one step ahead of the pathogens is not enough. Pathogen populations are often so large that subpopulations have already moved that first step toward resistance. A third problem is that as a community our antibiotic treatment protocols are not set to restrict the emergence of resistance. Consequently, it is unlikely that new antibiotics will solve the resistance problem without major changes in our philosophy about antibiotic use.

Overuse

One change would be to correct policies that enable the selective enrichment of resistant subpopulations. The medical community is in general agreement that we use many antibiotics when they are not needed. Excess prescription writing and self-medication are medical aspects of overuse that can be controlled. For example, half of the 100 million annual prescriptions in the United States for respiratory infections may be unnecessary. Part of the effort needs to be education of medical professionals at the earliest stages of their training. Another part is education of the public concerning the dangers of poor compliance and improper use of antibiotics. The educational message can be reinforced by legislation and enforcement in cases where antibiotics are sold outside the prescribing process. Reducing the abuse of prescriptions will probably become easier as the potential harm to the individual consumer becomes better documented (refer to Table 12-5): Less pressure placed on doctors to prescribe will reduce the number of prescriptions. But cultural issues are difficult to overcome with education when grocery stores offer free antibiotics.

The agricultural community must also do its share. Use of growth promoters and massive drug treatment to combat disease stemming from overcrowding apply selective pressure to a vast community of microbes that reaches the human food supply. Indeed, agricultural use of antibiotics is so much larger than medical use that one could argue that most of the emphasis on resistance should be directed at agriculture. Astute farmers and managers of agribusinesses will see that the days of massive antibiotic use are limited, and they will begin shifting to other strategies.

The argument for limiting use is also clear at the molecular level. Efflux mutations are readily obtained that lower susceptibility to many agents simultaneously (refer to Table 12-3); consequently, use of one type of antibiotic can start others on the climb to resistance. Moreover, antibiotic use favors cells containing integrons that have assembled resistance cassettes: Environmental contamination with antibiotics selects resistant organisms that can spread resistance genes to pathogens. Thus, an effort is required to lower environmental levels of antibiotics by controlling use and disposal at all levels.

Dosing

Changing dosing concepts is also important. We argued that when antibiotics are needed, doses are too low. Placing drug concentrations inside the mutant selection window allows selective amplification of mutant subpopulations. The suggestion that higher doses be used is often countered by the question, "Are the higher doses safe?" Safety issues are so firmly entrenched in the medical community and in the minds of the public that they override all other considerations. Indeed, the Food and Drug Administration was founded in the 1930s to protect the public from unscrupulous medicine men and their snake oils. There is no doubt that safety needs to remain a critical factor for drug development; however, safety issues need to be re-evaluated. Higher doses for many antibiotics would help limit the emergence of resistance. Some antibiotics can be used safely at higher levels. For example, penicillin doses have been increased substantially over time, and recently the dose of levofloxacin, a fluoroquinolone, was increased by 50% for some indications.

Drug Discovery and Surveillance

From an industry perspective, the question of resistance needs to enter the drug discovery process at an early stage. Drug resistance discussions currently come into serious play after a drug is introduced into the market. We argue that the criteria for developing new compounds should make prevention of resistance equal to considerations of safety and efficacy. For new drugs, a combination of creative chemistries and new performance criteria should lead to better antibiotics that last longer in the clinic. Indeed, in the future, only compounds that seriously restrict the emergence of resistance will experience widespread use because the others will be held back for "emergencies."

Some hope for short-term, local solutions can be seen in the aggressive effort being mounted with MRSA in several small European countries. The effort is best described by the "search and destroy" policy that Dutch and Danish health authorities have taken against MRSA. In Denmark, all MRSA-positive persons are offered eradication treatment, and guidelines recommend that they be issued personal MRSA identification cards. These cards must be shown at each contact with healthcare workers. Moreover, physicians are required to report all MRSA cases.[298] Such policies appear to be keeping the prevalence of hospital-associated MRSA low. Whether these policies will continue to contain the MRSA problem is uncertain, because a large reservoir is being generated in farm animals. Also unknown is whether aggressive policies will work in larger countries that have a higher prevalence of drug resistance or with diseases spread by routes other than direct contact. We are encouraged that many states in the United States now require reporting of pathogens such as MRSA. Moreover, individual institutions are beginning to assess the value of patient and healthcare worker decolonization strategies to prevent infection.

We still place hope in basic research. For example, we know that a compromised immune system can contribute to emergence of drug resistance. But is the effect simply due to higher levels of infecting organisms, or are other factors at play? Might human hormones, produced during stress, facilitate pathogen growth?[299] Those hormones could be manipulated. On the microbial side, we now realize the need to have a much better understanding of integrons. These DNA elements gather large numbers of resistance genes into a single DNA locale from which they can move to other bacteria. That makes the recipient microbe multidrug resistant in a single step. We have no way to stop this process. Moreover, it is unlikely that we could put together combinations of drugs in a way that would provide a long-term control over integron-mediated resistance. Perhaps we will find small-molecule inhibitors that will block the action of the integron integrase enzymes.

Resistance as a Side Effect

We close *Antibiotic Resistance* by reiterating that we collectively created a resistance problem that cannot be easily corrected. A major flaw in our approach has been to treat antibiotics like other drugs, to assume that side effects are the main features to guard against rather than resistance. By ignoring pathogen evolution, we lost control over malaria and pneumonia caused by *Acinetobacter*; we may soon lose the battle with gonorrhea. Perhaps a way to move forward is to emphasize that resistance is a harmful side effect of antibiotic use when that use predisposes us to future resistance problems.

Appendix A

Molecules of Life

To make *Antibiotic Resistance* accessible to a larger audience, we provide two appendixes that describe basic features of microbiology. In these appendixes, we assume that the reader has no background in biology; thus, we begin at a more basic level than is commonly seen in the popular press.

The Action of Molecules Defines Life

Many of the details of life can be understood by considering the behavior of molecules. For example, antibiotics are molecules, and their targets are molecules; antibiotic resistance arises through changes in molecules. Because atoms are the building blocks of molecules, our discussion of molecules must begin with atoms.

Atoms are tiny particles that exist in about 100 different types called elements. Familiar elements are oxygen, hydrogen, carbon, iron, and zinc. Each atom contains a nucleus composed of smaller particles called protons and neutrons. The number of protons in the nucleus defines the element. For example, hydrogen atoms have one proton; oxygen has sixteen. Each atom also contains a surrounding cloud of electrons. Protons behave as if they are positively charged, whereas electrons are considered to be negatively charged. An atom in its elemental form contains the same number of protons and electrons. Such an atom is electrically neutral.

Atoms can bond tightly to other atoms by sharing electrons. Such an interaction is called a covalent bond. Forming and breaking covalent bonds is called a chemical reaction; atoms bonded covalently act as a unit called a molecule. For example, a water molecule is two hydrogen atoms bonded to one oxygen atom (H_2O). The properties of water molecules differ from those of their components, hydrogen and oxygen. (At normal temperature and pressure, both hydrogen and oxygen are gases.) Molecules can be small, such as when two hydrogen atoms are bonded to form hydrogen gas, or they can be large, as with DNA molecules that can contain many billions of atoms. The key idea is that molecules are distinct entities that become larger or smaller through chemical reactions (forming or breaking covalent bonds between some of the atoms).

Atoms can have an electrical charge due to an imbalance between the number of protons and electrons. Atoms sometimes lose one or more electrons, which causes them to have a net positive charge. Atoms can also gain one or more electrons and take on a net negative charge. These charged atoms are called ions. Opposite charges attract, whereas like charges repel. These are called ionic interactions. Ionic interactions are weak when compared to covalent bonds. However, when many ionic interactions are involved as a collective, they can be strong. Ionic interactions are important for enabling some giant molecules to stick together and for forcing others to come apart. Ionic interactions also contribute to specific folding seen in large molecules.

The formation and destruction of biological molecules, the chemistry of life, is controlled by specialized molecules called enzymes. Enzymes serve as molecular catalysts: They accelerate chemical reactions without themselves being chemically changed. (Enzymes are often recognized by the suffix –ase). Some enzymes provide a surface or a pocket in which two other molecules dock in such a way that they become structurally distorted. The distortions can enable the formation of a covalent bond that combines the two molecules into a new, larger molecule. Other enzymes cause particular large molecules to break into smaller ones. Enzymes also carry out complex mechanical processes. For example, one of the enzymes that serves as an antibiotic target breaks DNA, pulls the ends apart, passes another region of the DNA through the break, and then closes the break. In a general sense, enzymes control the flow of chemical reactions. Defects in some enzymes cause fatal genetic diseases in humans. Likewise, antibiotics that disable an enzyme can cause a microbe to die.

Large biological molecules are built by joining many smaller molecules, somewhat like making a chain by adding links. The "links" are called subunits or monomers, and the long "chains" are called polymers. Formation of both the subunits and the polymers is controlled by specific enzymes designed for each task. Large biological molecules, such as DNA, RNA, and protein, are called macromolecules. Several macromolecules are discussed in the following sections because they play key roles in antibiotic action and resistance.

Proteins Are Molecular Workers

Most enzymes are a type of polymeric molecule called protein. Proteins are chains of amino acids. Twenty different amino acid types are found in living cells. When joined in a protein, the amino acids interact. Particular amino acids attract; others repel. The order of amino acids in a protein chain and the length of the chain determine the three-dimensional structure of a given protein molecule. That structure then determines how the protein functions in the cell. Differences among organisms can be reduced to differences in their proteins. Because protein length can range from less than a hundred to more than several thousand amino acids, an astronomical number of different protein molecules can exist. (For example, a chain of 100 amino acids could come in 20^{100} combinations—20 times 20 one hundred times). That gives life forms tremendous potential for diversity.

Proteins form many components of the cell, including the cables that provide structure to the cell, the machines that make new proteins and other large molecules, and the channels that allow molecules to pass into and out of cells. Many toxins (poisons) produced by pathogens and snakes are also proteins. Some of these toxins are enzymes that break down important molecules in our bodies. Others block the capability to make new molecules.

Proteins can be envisioned as the molecules that do the work in the cell and give the cell much of its form. Not surprisingly, proteins are the molecular targets of most antibiotics. One general class of antibiotic resistance arises from the protein target changing its structure such that the antibiotic no longer binds to the target.

DNA Is the Repository of Genetic Information

The instructions for constructing proteins is contained in extremely long, double-stranded molecules called DNA (deoxyribonucleic acid). DNA molecules are composed of a linear array of subunits called nucleotides. How information is stored can be understood by likening DNA to a book (see Box A-1). The nucleotides are arranged in sets called genes. In general, one gene contains the information for making one protein. DNA molecules are long enough to contain genes for the many thousands of different proteins that form and run cells.

Every organism on earth contains its own DNA molecules. It is the nucleotide sequence, in some cases billions of nucleotides long, that determines the properties of each organism. Changes in the nucleotide sequence of a gene can change the amino acid sequence in the protein specified by the gene. If that protein is involved in antibiotic action, the DNA change may cause antibiotic resistance. Thus, changes in the information in DNA are the molecular basis underlying the emergence of antibiotic resistance. A microbe or virus with such a change is called an antibiotic-resistant mutant.

Box A-1: Comparison of DNA and Books

The information in both DNA and books is stored in the order of "letters." DNA contains four letters, the four nucleotides abbreviated as *A, T, C*, and *G*; books written in the English language are composed of 26 letters. In both cases, the letters are arranged into words. In DNA, all words, which are called codons, are 3-letters long. Because DNA has only 4 different letters, it can have only 64 different words (4 x 4 x 4 = 64). In both DNA and books, the words are treated as groups. In DNA, a group of codons is called a gene; in books, the word groups are called sentences. A gene contains hundreds to thousands of codons; rarely does a sentence contain more than 20 words. Both genes and sentences are read through the linear order of either the codons in the case of DNA or words in the case of books. Each gene has the information that tells the cell how to make a specific protein, that is, the information in the DNA specifies the exact amino acid sequence and length of a particular protein. Many organisms contain several or even many different DNA molecules, that is, multiple chromosomes. This arrangement is comparable to the information held in a set of different volumes of books.

The analogy between DNA and books is imperfect because the story in a book is told by the linear order of sentences, whereas the story held in DNA is not told strictly by the order of the genes. Instead, the information in DNA can be accessed from many different sections, usually from many sections simultaneously.

When cells divide to form new cells, each new daughter cell must obtain a copy of DNA from the parental cell. That means that the information in DNA must be copied before cell division can occur. The double-stranded nature of DNA is the key to this process. The first point to note is that the two strands of DNA are not identical. Instead, they are complementary: An *A* nucleotide in one strand always corresponds to a *T* in the opposite strand, and a *G* nucleotide in one strand always corresponds to a *C* nucleotide in the other strand. The interaction of a nucleotide in one strand with a nucleotide in the other is called base pairing. In a process termed DNA replication, the two DNA strands are forced apart, and two new complementary strands are made, one paired with each old strand. Two double-stranded molecules result that are both identical to the starting DNA molecule. One double-strand DNA is passed to each daughter cell, thereby producing genetically identical daughter cells.

The pairing between complementary nucleotides is the most important concept of molecular biology, because it explains how genetic information is copied and how nucleic acids (DNA and RNA) recognize each other. Recognition is based on two strands of a nucleic acid, DNA or RNA, binding tightly to each other only when the fit between the strands is perfect or nearly so, that is, each *A* must align with a *T* or *U* (in RNA) and each *G* with a *C*. The principle is illustrated in Figure A-1.

Figure A-1 *Complementary base-pairing*

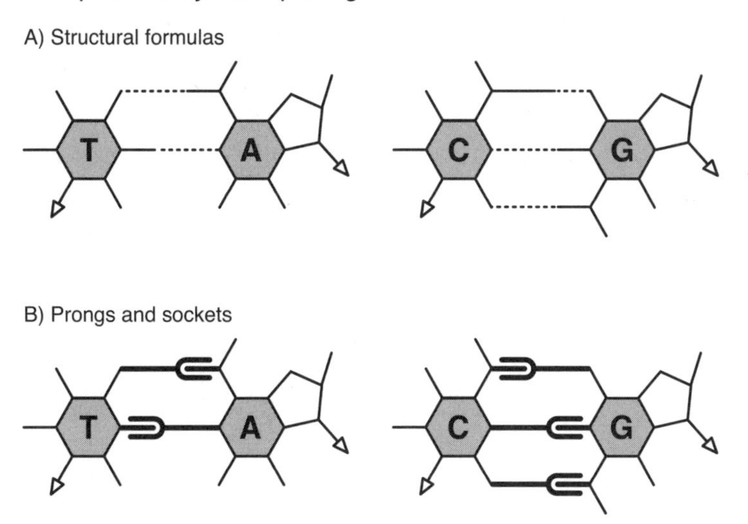

A) Structural formulas for thymine:adenine (*T*·*A*) and cytosine:guanine (*C*·*G*) base pairs. The bases are flat structures in which the solid lines represent chemical bonds between the atoms. Arrows indicate points at which the bases attach to sugars in DNA. The dotted lines are hydrogen bonds, weak attractive forces between hydrogen and either nitrogen or oxygen. (There are two hydrogen bonds between adenine and thymine, three between cytosine and guanine.) Differences in hydrogen bonding provides part of the explanation for complementary base pairing.

B) A prongs-and-sockets analogy for base pairing. The hydrogen atoms in each hydrogen bond are represented as prongs, and the oxygen or nitrogen atoms are depicted as sockets. The attractive forces are weak; consequently, perfect fits are required for base pairing to occur.

The machinery for making new DNA molecules is a group of proteins collectively called DNA polymerase. Occasionally, DNA polymerase makes an error—an incorrect nucleotide is placed in DNA when it is copied. That can cause a change in the protein specified by the gene in which the error occurred.

Because the error is in DNA, it will be passed from one generation to the next. We have no control over the generation of mutations except to make it worse. (In our daily lives, we encounter chemicals, called mutagens, that increase the frequency of DNA errors.)

The DNA strands are wound around each other like strands in a rope. Consequently, pulling the strands apart creates torsional tension. (Imagine pulling apart the strands of a two-stranded rope: Twisting will arise downstream from where the strands separate.) Enzymes called DNA helicases force the strands apart, and other enzymes called DNA topoisomerases relieve the tension. Together they enable DNA replication to proceed rapidly (800 base pairs per second). Antibiotics called fluoroquinolones interfere with topoisomerases and block DNA replication.

Although DNA molecules are stable enough to reliably pass genetic information from one generation to the next, they are not static (see Box A-2). One DNA activity important for antibiotic resistance is called genetic recombination. In this process, one DNA molecule exchanges a section with another DNA molecule by breaking and rejoining ends. Certain DNA molecules can also insert into and excise from others. Such events can have important implications for the evolution of all organisms, including pathogens.

Box A-2: DNA Is Dynamic

DNA molecules are flexible. Their physical flexibility is easy to understand when we consider how long and thin they are: If all the DNA molecules in one human body were lined up end to end, they would stretch to the sun and back 300 times. After years of study, we now know that DNA molecules can be tied into knots, bent into loops, coiled like a telephone cord, wrapped around proteins to form what look like balls, and linked/unlinked when circular (the magician's ring trick). Cellular proteins direct these manipulations of DNA. When considered over evolutionary time, we see that the information content (nucleotide sequence) of DNA is also dynamic. By examining nucleotide sequences, we can observe remnants of ancient insertions, deletions, and duplications of genetic material. When small DNA molecules move from one microbial cell to another, they can carry antibiotic resistance genes; thus, when we use antibiotics, we favor the movement of those genes because organisms that contain them will have a growth advantage over those that do not. We do not know how to stop this natural gene movement.

Every organism has its own DNA with its own unique nucleotide sequence; consequently, that sequence information can be used for identification of pathogens and, in some cases, for determination of antibiotic resistance. One of the problems with diagnosis is that the pathogen DNA may be present at low amounts. A method called the polymerase chain reaction (PCR) enables us to amplify short nucleotide sequences to obtain enough for analysis. (For an explanation of PCR, see Box A-3.)

Box A-3: Polymerase Chain Reaction

The polymerase chain reaction (PCR, see Figure A-2) makes it possible to selectively amplify any region of DNA as long as nucleotide sequences are known for short (15-nucleotide) regions on each side of the DNA stretch to be amplified. Several aspects of DNA replication are important for understanding PCR. First, new DNA is made by an enzyme called DNA polymerase. It travels along a single strand of DNA, making a second strand by adding nucleotides one at a time to a new, growing chain. The new DNA strand has a sequence of nucleotides complementary to that in the old, template strand (*T* in one strand is paired to *A* in the other, and *G* in one is paired to *C* in the other). As the new strand is made, it is bound to the old template strand by complementary base pairing. Second, each DNA strand has directionality, much like a line of elephants hooked head to tail. The polymerase recognizes that directionality and makes new DNA only in one direction. Third, in double-stranded DNA, the two strands run in opposite directions. When the two strands are separated, DNA polymerase will make new DNA from left to right with one template strand and from right to left with the other. Fourth, DNA polymerase requires a pre-existing end of DNA to make new DNA: It adds new nucleotides only to the *end* of a single strand of DNA. Consequently, a short single strand of DNA, hybridized to a long single strand, will create a starting point for DNA polymerase. The short single-strand DNA is called a primer. Two primers are chosen that hybridize to different regions of template DNA: one primer to one template strand and the second primer to the other complementary template strand. The primers are chosen so new DNA is made from the first primer toward the second (left to right) and from the second toward the first (right to left). Thus, new DNA is made from a specific region between (and including) the primers. The procedure follows:

1. Mix together DNA polymerase, template DNA, nucleotides, and the two primer DNAs.
2. Heat to cause the two strands of double-stranded template DNA to come apart.

3. Cool to allow primers to hybridize with the template strands and to allow polymerase action. Twenty to thirty cycles of heating, cooling, and polymerization cause a specific region of DNA to be copied over and over.

The polymerase chain reaction is so useful that its inventor, Kary Mullis, was awarded a Nobel prize.

Figure A-2 *Polymerase chain reaction*

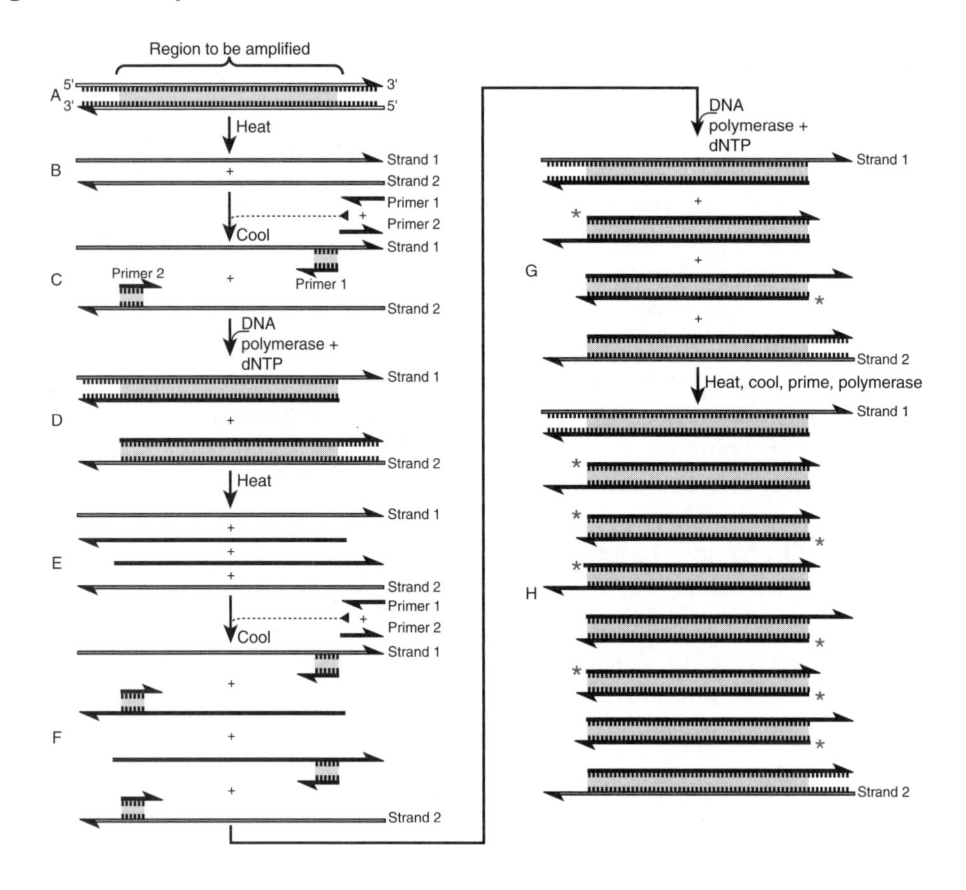

A A double-stranded DNA molecule that contains a specific region of interest. (Double-strandedness is indicated by short, vertical lines.)

B Heating the DNA causes the two strands to separate.

C When primers complementary to short regions on each of the two strands shown in A are present in the mixture and it is cooled, the primers hybridize to the two strands labeled 1 and 2.

D The primers are extended by inclusion of DNA polymerase and nucleoside triphosphates in the reaction mixture. The new DNA molecules (solid lines) have different lengths.

E When the mixture is heated, all the DNA molecules become single stranded.

F Upon cooling, primers hybridize to both old and new strands in the mixture.

G DNA polymerase again extends the primers. If the templates are the strands made in step D, DNA synthesis stops when the polymerase reaches the end of the template. This produces a discrete DNA fragment (*).

H A subsequent cycle of heating, cooling, and polymerization increases the relative abundance of the discrete fragment (*). Heat-resistant DNA polymerase is used. Therefore, if enough polymerase, nucleoside triphosphates, and primers are added at the beginning of the reaction, the process consists of only heating and cooling steps.

RNA Plays Several Roles in Life Processes

RNA (ribonucleic acid) molecules are similar to DNA, but they are shorter and usually single stranded rather than double stranded. The subunits (nucleotides) of RNA also have a slightly different structure than DNA nucleotides. Moreover, RNA has *U* nucleotides rather than *T*s. RNA strands recognize each other and DNA strands by complementary base pairing: Two single-stranded nucleic acids (DNA and RNA) that are complementary tend to stick together. Single-stranded RNA molecules fold into complex structures when distant, complementary regions form base pairs. The result is short, double-stranded sections separated by single-stranded loops.

Ancient life forms probably contained RNA but neither DNA nor protein. That period in Earth's history is sometimes called the RNA World, because RNA is likely to have directed life processes. Some RNA molecules still act as enzymes by accelerating chemical reactions. (Examples are RNA molecules that break other RNA molecules.) The existence of RNA viruses shows that RNA can also store genetic information and thus provide the functions of DNA. (Viruses such as influenza virus, poliovirus, and many plant viruses use RNA, not DNA, to store their genetic information.)

In modern cells, RNA molecules are made from information in DNA. (With many RNA viruses, which are not cells, RNA can be made from information in RNA.) In cells, specific RNA molecules take information from DNA and move it to workbenches (ribosomes) where the information is used to make proteins.

These RNA molecules are called messenger RNA (mRNA). Genes in DNA have start and stop signals that identify regions from which an RNA-synthesizing enzyme (RNA polymerase) makes RNA copies of the nucleotide sequence. (The process of making RNA molecules from information in DNA is called transcription.) Cells contain many RNA polymerase molecules that enable cells to make RNA copies from many genes at once. Thus, cells have access to huge amounts of information at short notice. They also have ways for deciding when the information in a particular gene will be accessed (see Box A-4). Some antibiotics, such as rifampicin, work by inactivating RNA polymerase, thereby preventing the formation of new RNA molecules.

Box A-4: Control of Gene Expression

Particular nucleotide sequences mark the beginnings and ends of genes, thereby signaling to RNA polymerase where to work. The beginning is called a promoter, which serves as a binding site for RNA polymerase. In some genes, particular protein molecules bind at or near the promoter and block binding by RNA polymerase. Such proteins are called repressors. Other proteins can act as activators that stimulate binding of RNA polymerase to the promoter of a gene. Repressors and activators are often sensitive to the environment, which enables the cell to change its protein composition to suit particular environmental conditions.

After the information in a gene is transcribed from DNA into mRNA, mRNA attaches to ribosomes. That effectively moves a portion of the information in DNA to the ribosome. In bacteria, ribosomes bind to mRNA while it is still being made from DNA; in higher cells, the mRNA releases from DNA in the nucleus and then travels to ribosomes in the cytoplasm. The mRNA binds to a reading site on the ribosome and begins to thread across it. At the same time, the ribosome constructs a new protein molecule based on the nucleotide sequence of the mRNA molecule.

Alignment of amino acids to form new protein occurs through small adapter RNA molecules called transfer RNA (tRNA). These RNA molecules, which are about 80 nucleotides long, fold into distinct structures. Each of the 20 different amino acids is assigned a different tRNA and is attached to one end of its cognate tRNA. Another part of the tRNA molecule recognizes a specific nucleotide triplet (three adjacent nucleotides) in mRNA. These triplets of nucleotides in mRNA and DNA are the genetic words called codons; the recognition region in the tRNA is called an anticodon, because it is complementary to the codon. When mRNA is situated in the reading site on the ribosome, its start codon is exposed for interaction with one particular type of tRNA anticodon. Only the tRNA whose anticodon is complementary to the mRNA start codon forms base pairs and binds tightly. The second codon of the mRNA is also exposed on the ribosome. A second tRNA, with its attached amino acid, binds to the second codon on mRNA. That brings the first and second amino acids close together. The ribosome then stimulates covalent bond formation between the two adjacent amino acids, thereby beginning the new protein chain. The first amino acid of the new protein separates from its tRNA, the mRNA molecule shifts (moves) one position on the ribosome, and the first tRNA detaches from the ribosome. The shift of mRNA moves the start codon out of the reading site and moves the third codon in. A third tRNA, with its attached amino acid, enters the reading site and binds to the third codon of mRNA. That places the third amino acid close to the second, and the two are covalently joined. At that point, the growing protein chain is three amino acids long. The second amino acid then releases from its tRNA, which exits the ribosome. This process of protein synthesis continues as the mRNA moves over the ribosome, much like an audio tape over the player head of a cassette player. Protein synthesis is called translation because information encoded in the 4-letter alphabet of nucleotides is converted into the 20-letter amino acid language of proteins. Eventually, perhaps after the information in a thousand nucleotides has been translated, a stop signal is reached. Then, the new protein is released from the ribosome, and it begins to do its job in the cell. Most of our antibiotics work by interfering with particular steps in protein synthesis.

Three-nucleotide codons are used to specify each amino acid. (The assignment of specific amino acids to specific nucleotide triplets is called the genetic code.) DNA and RNA have 64 different codons (4 nucleotides taken in sets of 3 = 4 x 4 x 4 = 64). Because cells have only 20 different amino acid types, many amino acids have more than one corresponding codon and more than one cognate tRNA species. This redundancy provides genetic flexibility.

Ribosomes have been an essential part of life for a long time. As organisms evolve into new species, the ribosomes also evolve to enable the organism to compete with other life forms. As a result of these changes, some bacteria are more sensitive than others to particular antibiotics that function by attacking ribosomes. The differences in ribosomes also provide a way to classify organisms and even a way to detect bacteria that we cannot grow in culture (see Box A-5).

Box A-5: Ribosomal RNA Is Used to Analyze Populations of Microbes

Ribosomes contain several large RNA molecules. The nucleotide sequence of a portion of the 16S rRNA is commonly used for analysis of bacterial populations. (Historically, RNA molecules were named according to their size, usually with a number followed by the letter *S*, which represents how rapidly the molecule sediments during centrifugation.) All cellular organisms contain ribosomes that are remarkably similar, because the complicated process of protein synthesis has remained essentially the same for billions of years. However, small changes in the nucleotide sequence of ribosomal RNA have accumulated over the years and can now be used to identify groups of bacteria. Methods are available to perform the analysis without growing the cells. From these analyses, we learn that many more types of bacteria exist than we can grow in culture. We can examine 16S rRNA sequences from skin bacteria or stomach contents to determine how many different types of bacteria are present and how those types change when we perturb our bodies with antibiotics.

Carbohydrates Store Energy, Form Cell Walls, and Make Rigid Structures

Carbohydrates include sugars, starch, and cellulose. They make up the woody parts of plants and the hard exoskeletons of insects. They are also important components of the cell walls of bacteria and fungi. Simple sugars contain several carbon atoms (glucose has six), hydrogen atoms, and oxygen atoms. These sugars can be joined to form large polymers, such as starch and cellulose. Nucleic acids also contain sugars. For example, each nucleotide (subunit) of DNA and RNA contains a five-carbon sugar (deoxyribose and ribose, respectively). As with other complex polymers, cell walls are built by enzymes that control the process. Some antibiotics, such as penicillin, interfere with the placement of sugars in bacterial cell walls. As a result, bacteria treated with penicillin stop making cell-wall material, leaving holes in the wall. That often causes bacteria to lyse (break apart).

Sugars are also important for cellular energy. For example, plants convert the energy of sunlight into chemical energy by making simple sugars from carbon dioxide and water. These sugars then join together to make complex sugars, such as starch, for energy storage. When energy is needed, starch is broken down to simple sugars. Sugars are consumed for energy by other organisms and by the nongreen parts of plants. Because sugars are not easily used to drive chemical reactions, they are broken down in a process that makes adenosine triphosphate (ATP), the energy molecule used by most cells. ATP is then "burned" to drive chemical reactions in cells, often through coupling of ATP breakdown with enzyme action.

In our cells, sugars are also used to decorate proteins after the proteins are made. These decorations serve as address labels that direct particular proteins to certain parts of the cell and even to certain parts of the body. The surface proteins of viruses sometimes contain sugars that influence the capability of our immune systems to mount an effective antiviral defense.

Lipids Store Energy and Form Membranes

Fats and oils are called lipids. Chemically, they are often long chains of carbon atoms to which hydrogens are covalently bound. That leads to the name hydrocarbon. Fats and oils do not dissolve in water. Instead, they tend to clump and exclude water. (When you add vegetable oil to water and shake, you can see oil droplets suspended in the water.) For this reason, fats are said to be hydrophobic. Hydrophobic interactions play important roles in the structure of macromolecules (see Box A-6).

Box A-6: Hydrophobic Interactions

Some amino acids are hydrophobic. When a part of a protein, these amino acids tend to cluster on the inside of the protein in their attempt to avoid water. That pulls on the amino acid chain of the protein and contributes to the protein shape. Other amino acids are hydrophilic. They are readily soluble in water, and they tend to be on the outside of proteins. Hydrophobic interactions also affect the capability of molecules to come together. For example, enzymes are often composed of several separate proteins. The surfaces where the subunits of an enzyme touch are frequently composed of hydrophobic amino acids. Hydrophobic interactions also affect the structure of DNA and RNA. For example, a part of each nucleotide of DNA and RNA is a planar (flat) ring structure called a base. The bases are hydrophobic, and to escape water they stack. This stacking provides rigidity to DNA. The bases of separate strands are on the inside of the double helix due to their hydrophobic interactions; the hydrophilic phosphate groups are on the outside.

Cells are surrounded by a membrane (plasma membrane) composed of lipid molecules that have a negatively charged phosphate on one end. The phosphate ends are drawn toward water, and the lipid ends try to escape water. The result is a two-layered membrane in which the lipid portions are inside and the phosphates point out. Cells use protein channels that span membranes to enable certain molecules to enter the cell while keeping others out. Consequently, cell membranes are said to be semi-permeable. Membrane proteins also pump some molecules out. Any process that restricts entry of antibiotics or increases antibiotic efflux lowers drug susceptibility and contributes to resistance. Some drugs, such as azoles, alter fungal membrane properties, punching holes or changing the flow of vital nutrients. Azoles are widely used to treat fungal infections.

Cellular Chemistry Is Organized into Metabolic Pathways

In living systems, the forming and breaking of covalent chemical bonds occurs in an ordered way that can be described as a vast network of interactions. The network can be sorted into many sets of pathways that trace the steps taken to construct or break down particular molecules. Some key pathways are common to most organisms. An example is conversion of energy stored in glucose to a more usable form as ATP. Specialized pathways may be unique to certain types of organisms. The pathways are connected by regulatory molecules that can shift the utilization of pathways. For example, if oxygen is removed from a culture of bacteria, a new set of proteins is made to facilitate an alternative strategy for gaining energy. When bacteria experience stress from antibiotic action, a protective set of pathways is used to survive the stress. Suicide pathways also exist. Some are important for proper development of multicellular organisms, whereas others protect from pathogen attack. Even bacteria are sometimes thought to have suicide pathways.

We now know the entire nucleotide sequence for the DNA of many pathogens. In principle, that should reveal everything about them. However, we don't know the meaning of all the sequence information. We are far from understanding the interactions among the metabolic pathways, how they communicate with each other, and how the pathways mount a coordinated defense against our effort to poison them.

Appendix B

Microbial Life Forms

Bacteria Lack Nuclei and Other Organelles

Bacteria cause some of our most notorious infectious diseases: plague, typhus, tuberculosis, and cholera are entries on a long list. These microbes are minute, single-celled organisms that contain all the information needed to reproduce. Bacteria reproduce by binary fission (splitting apart). DNA is duplicated and pulled to separate halves of the cell; then a ring forms in the cell wall, much like a ring on your finger. The ring gets tighter and tighter until it pinches the cell in two. After cell division, the two daughter cells grow until they, too, divide in half. Thus, the number of cells in a bacterial culture increases by doubling: 1, 2, 4, 8, 16, 32, 64, and so on. With some bacteria, the doubling time can be as short as 20 minutes, which enables a culture to go from one cell to a million in 7 hours. Such rapid growth makes it important to treat an infection promptly. Other bacteria grow more slowly (the tuberculosis bacillus takes 24 hours to double), but they can be just as deadly.

Bacteria are called prokaryotic organisms because their DNA is not packaged in a true nucleus, a microscopic membrane-bound structure seen in cells of higher organisms such as ourselves. (Cells with a nucleus are called eukaryotic.) This difference between prokaryotic and eukaryotic organisms reflects evolutionary paths that separated roughly two billion years ago. Over that time even highly conserved aspects, such as the machinery for making proteins, changed enough for antibiotics to block bacterial growth with little effect on our cells.

Most bacterial species are not pathogens. When we treat with antibiotics for a harmful bacterium, we also eliminate helpful ones. The resulting disturbance of the microbial ecosystem in our bodies can enable other harmful microbes to take over. Treatment with antibiotics has risk.

Fungi Are Eukaryotes Having Cell Walls But Not Chloropasts

Yeasts, molds, and mushrooms are fungi. Unlike bacteria, fungal cells store their DNA in true nuclei. Such subcellular structures, called organelles, localize particular cellular functions. For example, mitochondria are power plants that convert chemical energy from sugars into molecules such as ATP. Lysosomes are the cellular equivalent of garbage disposal units. They are filled with enzymes that destroy other macromolecules. Bacterial cells lack such localization of cellular activity.

In general, the molecules that constitute fungal cells are similar to those of human cells. Consequently, antibiotics that attack bacteria usually fail to affect fungi, largely because the agents were designed to have little activity against eukaryotic organisms. However, fungal cells do differ from our cells in fundamental ways that can be exploited. One of those differences involves the structure of the cell membrane. Most clinical antifungal agents target enzymes that participate in making components of the membrane.

Yeasts are single-celled fungi that grow much like bacteria. Liquid cultures become cloudy when many yeast cells are present, and yeast form colonies on agar. Some yeast species reproduce by dividing (fission yeast), whereas others form buds that break off and grow into new cells. (A bud is a protrusion of the cell that gradually increases in size until it pinches off.) Bakers' yeast, commonly used to make bread, is one of the most thoroughly studied eukaryotic organisms. Its cousin, *Candida albicans*, kills immune-deficient persons.

We often think of molds as the fuzzy green or black growth on old bread. Molds are multicellular organisms whose bodies consist of long, thin filaments called hyphae. Some hyphal cells develop into fruiting bodies full of tiny spores. The spores are released into the air; by breathing, we draw them deep into our lungs. There, they germinate and sprout hyphae that penetrate our lung tissue. Some filamentous fungi convert to a yeast form in our lungs, a conversion crucial for their invasion of our lung tissue (see Box B-1).

Yeasts and molds are everywhere in our environment. For most persons, they are not a problem because healthy immune systems remove fungal cells from our bodies. However, as our populations age and immunosuppression becomes increasingly common, fungal diseases also become increasingly common. Invasive fungal infections are now a major cause of death for cancer patients. In some types of patient, fungal infections account for nearly half the deaths. Newborns are also susceptible to yeast infection, because their immune systems are poorly formed.

Box B-1: Molds That Become Yeasts

Some pathogenic fungi normally live in the environment, often in soil, as filamentous mycelia that produce spores. Environmental disruption (plowing and earthquakes) can aerosolize mycelial fragments and spores, enabling them to be easily inhaled by humans. Inside the lungs, the fungi cause a disease, valley fever, that closely resembles pneumonia. In the southwestern United States, fungal disease accounts for 15% to 30% of community-associated pneumonia. If these cases are misdiagnosed as bacterial pneumonia, treatment will be ineffective because antibacterial agents fail to block fungal growth (www.cdc.gov/mmwr/preview/mmwrhtml/00025779.htm).[300]

When the fungi are in the lungs, they change to a yeast form (round, ball-like structures). If this switch is not made, the host immune system clears the fungus from the body. A warming to body temperature is important for the switch, as is carbon dioxide concentration. (Carbon dioxide is 150 times more concentrated in the lung than in the atmosphere.) With one of these dimorphic fungi, *Paracoccidiodes brasiliensis*, the switch to the yeast form is blocked by estradiol, a steroid found in human females. Before puberty, boys and girls are equally susceptible to the fungal disease. After puberty, the fungus victims are overwhelmingly boys. In general, the severity of infection increases with fungus dose and conditions such as AIDS, diabetes, and pregnancy.[301]

Parasitic Protozoa Are Eukaryotes Lacking a Cell Wall

Protozoa are small, single-celled organisms found in many environments. Several species parasitize humans, causing serious diseases such as malaria and sleeping sickness. Inside human hosts, parasitic protozoa often have complex life cycles, with distinct forms in liver and blood. Some of these protozoa have still another distinct form when inside an insect, usually a mosquito or biting fly. This is the stage that spreads when a person is bitten by the insect. Malaria, which is caused by four distinct parasite species, is a major killer in countries unable to control mosquitoes (the World Health Organization estimated 246 million malaria cases in 2006 with almost 900,000 deaths). Malaria is such a devastating disease that humans in Africa and the Mediterranean basin have acquired mutations, such as sickle-cell trait, that provide partial protection. (Sickle-cell trait is a genetic condition in which one of the two copies of a hemoglobin gene is mutant; when both copies are mutant, the affected person suffers from sickle-cell disease.)

Helminths Are Parasitic Worms

Parasitic worms (helminths) are multicellular organisms that live inside humans and animals. Worm offspring are passed to human hosts through poorly cooked meat, contaminated water, and mosquitoes. Toxins produced by the worms can increase host susceptibility to a variety of other infections. While diseases caused by parasitic worms are rare in the U.S. and other industrialized countries, they commonly result in blindness and dysentery in developing countries. Pinworms are the most common helminth infection; in the U.S. they frequent travelers, migrant laborers, and the homeless.

Viruses Are Inert Until They Infect

In molecular terms, viruses are much simpler than cellular pathogens and parasites. They are composed of a nucleic acid, either RNA or DNA, a protein coat that protects the nucleic acid, and often several other proteins that form an outer envelope. Some viruses also carry important enzymes with them. The protein coat of viruses can be simple, or it can comprise many proteins that form the complex head and tail structures seen with some bacteriophages (viruses that attack bacteria). Other viruses, such as human immunodeficiency virus (HIV), acquire an outer coat of membrane when they exit human cells (release of virus occurs by the infected cells forming small buds that pinch off, releasing membrane-coated virus). We stress that no virus makes its own proteins; consequently, all viruses seek out living cells and force them make viral proteins and new copies of viral nucleic acids. In some cases, the viruses take the host protein synthesis machinery for their own exclusive use. In other cases, the host cell is allowed to continue making its own proteins.

Some viruses kill their host cells outright, while others have the option of entering a dormant state. That dormant state can involve insertion of viral DNA into the host chromosome. When conditions change, the virus can literally pop its DNA out of the host chromosome. Such occurs with some bacterial viruses. After the viral DNA excises from the chromosome, it duplicates, viral proteins are produced that combine with the DNA to make new infectious virus particles, and then other viral proteins break open the host cell. HIV enters its host cell as an RNA virus, converts its genetic information to a DNA form, and then inserts that DNA into a host chromosome. HIV DNA stays in the host chromosome until the host dies. From its chromosomal location, HIV DNA directs formation of new virus parts, including copies of viral RNA for new virus particles. In each of these scenarios, viral genes encode viral proteins that enable the virus to reproduce.

Glossary

agar: An inert gelatin-like substance that provides a solid surface for growth of microbial colonies.

amino acid: One of 20 small molecules that comprise the subunits of proteins.

amino acyl tRNA synthetase: An enzyme that attaches a specific type of amino acid to a specific type of transfer RNA in preparation for protein synthesis; each amino acid-tRNA combination uses a different amino acyl tRNA synthetase.

antibiotic: A chemical that controls infection by blocking microbial growth and, in some cases, by killing microbes while doing little harm to the mammalian host of the microbes; in *Antibiotic Resistance*, antibiotics are considered to include antibacterial, antiviral, antifungal, antiprotozoan, and antihelminth agents.

antibiotic resistance: The capability of a particular pathogen population to grow in the presence of a given antibiotic when the antibiotic is used according to a particular regimen. See clinical resistance.

antigen: A chemical group recognized by an antibody.

antisense oligonucleotides: Short single-stranded DNA molecules designed to hybridize to specific nucleic acid targets.

antiseptic: A chemical that kills microbes but is safe enough to be applied to human skin.

area under the curve (AUC): A measure of antibiotic concentration over time; determined by plotting concentration versus time and then measuring the area under that plot.

bacteremia: Blood infection.

bacteriophage: Virus that infects bacteria.

bacterium: A single-celled organism that reproduces by binary fission and lacks a true nucleus, mitochondrion, or chloroplast.

base pair: Two nucleotides, each in a different strand of DNA, that interact; generally as complementary base pairs such that A pairs with T or U and G pairs with C.

bioavailability: The degree to which or rate at which an antibiotic is absorbed or becomes available at the site of physiological activity.

biofilm: A complex aggregation of microbes growing on a solid surface, often exhibiting reduced susceptibility to antibiotics.

breakpoint: (interpretative breakpoint) A value of MIC that correlates with antibiotic failure; when an isolate has an MIC above the breakpoint, it is considered to be resistant.

broad-spectrum antibiotic: Having activity against many pathogen species.

CA-MRSA: Community-associated, methicillin-resistant *Staphylococcus aureus*, a common cause of skin and soft-tissue infections outside hospitals or other institutions.

cassette integration: A process by which genes are moved and assembled by an integron.

chain termination: A process in which incorporation of an "abnormal" nucleoside into DNA blocks further DNA synthesis; many antiviral agents act by chain termination.

chromosome: Large DNA molecule that contains nucleotide sequence information specifying properties of an organism.

clinical resistance: A situation in which the MIC of an isolate is above the interpretative breakpoint.

clone (noun): A group of cells that have all descended from the same cell.

codon: Nucleotide triplet in DNA and mRNA that corresponds to a specific amino acid in protein.

codon usage: The bias exhibited by an organism for use of a particular set of tRNAs and their corresponding codons in protein synthesis; two regions of DNA having different codon usage are thought to have different evolutionary origins.

colony: A visible growth on a solid surface arising from multiple rounds of microbial reproduction.

commensal: An organism benefited by the presence of another organism without affecting the second organism.

complementary base pair: Two nucleotides, each in a different strand of DNA, that interact such that A pairs with T or U and G pairs with C.

confluent growth: A situation occurring with cultured animal cells in which the cells stop growing and dividing when they touch each other; the result is a layer of cells one-cell thick on the bottom of a Petri dish.

conjugation: The process by which DNA is actively passed from one microbial cell to another; with bacteria, conjugation generally involves a plasmid.

cross resistance: A resistance mechanism that applies to two or more antibiotics; an example is drug efflux in which resistance is conferred to multiple antibiotics.

cytokine: A member of a family of polypeptide regulators of the immune response; an example is interferon.

directly observed therapy (DOT): A procedure in which healthcare workers visit tuberculosis patients to assure that the patients take the prescribed antibiotics.

disinfectant: A chemical that kills microbes and is generally applied to inanimate surfaces.

DNA: Deoxyribonucleic acid; a long, thin, chain-like molecule that contains the information for nucleotide sequences of RNA molecules and amino acid sequences of proteins.

DNA duplex: Double-strand DNA.

DNA polymerase: An enzyme complex that makes new DNA molecules by adding nucleotides one at a time to the end of a growing DNA molecule using a complementary DNA strand as a template.

DNA replication: Synthesis of DNA.

DNA virus: A virus that uses DNA as its genetic material.

efflux pump: A set of proteins that work together to expel antibiotics and other small molecules from cells.

empiric therapy: Use of antibiotics to treat disease without laboratory tests to identify the caustive pathogen or to determine its antibiotic susceptibility.

enzyme: A macromolecule, usually protein, that accelerates a chemical reaction without being consumed by the reaction.

eukaryotic: A type of organism that has a true nucleus and generally other organelles.

extensively drug resistant (XDR): A strain of pathogen resistant to most available antibiotics; with *M. tuberculosis*, XDR means resistance to isoniazid, rifampicin, a fluoroquinolone, and one injectable drug (capreomycin, amikacin, or kanamycin).

fitness: A measure of the capability of an organism to grow and survive in a particular environment.

focus (noun): A pile of animal cells generated by infection with a tumor virus that prevents the cells from ceasing growth when they touch other cells.

fungus: A type of eukaryotic organism characterized by having a cell wall but lacking chloroplasts; examples are yeast, molds, and mushrooms.

gene mobilization: A process by which a chromosomal gene is transferred to a plasmid that then transfers the gene to other bacterial cells.

general recombination: See homologous recombination.

generalized transduction: Movement of bacterial genes from one bacterial cell to another by a bacteriophage such that any gene can be moved.

genome: The information content (nucleotide sequence) of an organism.

genomic island: A region of nucleotides in the genome of an organism in which the AT to GC ratio and/or codon usage in the region is different from the bulk of the genome, suggesting that the region of nucleotides had an origin different from the bulk of the genome.

growth promoter: Antibiotic administered at low doses to food animals to enhance animal growth.

HA-MRSA: Hospital-associated methicillin-resistant *Staphylococcus aureus*, a common cause of serious infections in hospitals and other institutions.

helminth: A type of worm.

high throughput screening: A process in which large numbers of molecules are tested for activity, generally using robotics for sample testing.

homologous recombination: Breakage and rejoining of DNA regions of similar (homologous) nucleotide sequence such that two DNA molecules exchange regions.

horizontal transfer: Movement of genes from one microbial cell to another, as opposed to inheritance of genes from mother cell to daughter cell.

hydroxyl radical: A small, highly reactive molecule composed of an oxygen, a hydrogen, and an excess electron (neutral form of hydroxyl ion); considered to be a reactive oxygen species.

inflammation: A complex response of a host to injury or infection in which leukocytes and plasma move from the blood to the injured region.

insertion site: Location in a DNA molecule where another DNA molecule inserts (integrates); often recognized by a specific protein (integrase).

integrase: A protein that facilitates the insertion (integration) of one DNA molecule into another.

integration: A process in which one DNA molecule inserts into another.

integron: A region of DNA that contains an integration site, a gene encoding an integrase, and a promoter such that genes from other regions of DNA are inserted downstream from the promoter, assuring their capability to be expressed.

interference RNA: Small RNA molecules that hybridize to complementary RNA molecules, forming a hybrid that is then cleaved by an enzyme that recognizes double-strand RNA; used to selectively inactivate expression of particular genes.

interpretative breakpoint: See breakpoint.

isolate (noun): A pathogen sample taken from a patient.

isolate (verb): To separate a microbe or a particular type of molecule from other "contaminating" microbes or molecules.

hybridization: See nucleic acid hybridization.

latent infection: A situation in which a pathogen is present but undetectable by direct means.

lawn: A growth of bacterial cells that completely covers the surface of an agar plate.

longitudinal surveillance: Survey of antibiotic resistance conducted periodically, often annually, in the same patient population and geographic region to provide information about changes in the prevalence of resistance.

lysis: Break apart, commonly referred to host cells breaking apart during virus infection.

lysogeny: The process by which a bacterial virus integrates into the bacterial chromosome and remains quiescent.

macrophage: A type of mammalian cell that acts as a scavenger of foreign particles and microbes.

maximal concentration (C_{max}): The highest level achieved by an antibiotic after administration.

messenger RNA (mRNA): A type of RNA that carries information from a gene in DNA to ribosomes where new protein is made using the nucleotide sequence in the mRNA to specify the number and order of amino acids in the new protein.

microbes: Cellular micro-organisms that include bacteria, yeasts, molds, and protozoa.

microbiome: The set of microbes carried by a multicellular host.

minimal bactericidal concentration (MBC): The antibiotic concentration that causes a 99.9% decrease in microbial survival in a specified time, usually 18 hours for rapidly growing bacteria.

minimal inhibitory concentration (MIC): Lowest concentration of an antibiotic that blocks the growth of a standard microbial inoculum, usually between 10^4 and 10^5 cells in the case of bacteria.

MRSA: Methicillin-resistant *Staphylococcus aureus*; a common commensal skin bacterium that can cause serious infections; commonly found on skin and inside the nose.

multi-drug resistant (MDR): A microbe or virus resistant to more than one antibiotic; in the case of *M. tuberculosis*, MDR refers to bacteria resistant to isoniazid and rifampicin.

mutant: An organism or virus that has undergone a change in its genetic material (DNA or RNA) that causes the organism or virus to differ from the parental population (not a recombinant).

mutant prevention concentration (MPC): Minimal concentration that blocks the growth of the least susceptible, next-step resistant mutant subpopulation; with bacteria MPC is approximated by the concentration that prevents colony formation when more than 10^{10} bacteria are applied to drug-containing agar.

mutant selection window: See mutant selection window hypothesis.

mutant selection window hypothesis: The proposition that antibiotic-resistant microbes and viruses are selectively enriched and selectively amplified when antibiotic concentrations fall in a specific range (mutant selection window); the lower boundary of the range is approximated by MIC and the upper boundary by MPC.

mutation: A change in the genetic material (DNA or RNA) of an organism or virus; mutations are commonly insertions, substitutions, and deletions of nucleotides.

narrow-spectrum antibiotic: Having activity against only a few species.

neuraminidase: An enzyme on the surface of influenza virus that removes sugar groups from proteins and lipids; a component of influenza virus.

nonadherence to therapy: Failure of a patient to comply with a therapeutic regimen.

nucleic acid: DNA and RNA.

nucleic acid hybridization: A process in which two single-stranded nucleic acids from different sources form a double-stranded molecule due to complementary base pairing.

nucleoside analogues: A nucleoside similar to A, T, G, C, or U that can be incorporated into DNA or RNA but does not function like a natural nucleoside.

open reading frame (orf): A region of DNA or RNA that can encode a protein; a region of nucleotide sequence generally devoid of stop codons.

opportunistic infection: Growth of a microbe that normally does not occur with a healthy host due to action of host immune system; common examples are fungal infections of immunocompromised individuals.

pandemic: A disease outbreak that spreads to multiple countries.

pasteurization: Heating a liquid, such as milk, to slow spoilage; in practice, the heating eliminates 99.999% of the viable microorganisms.

pathogenicity island: A region of DNA in a microbial chromosome that 1) contains genes that contribute to pathogenicity and 2) is thought to have moved into the microbe as a block from another organism.

peroxide: A small oxygen-containing molecule of the form H_2O_2 (hydrogen peroxide); considered to be a reactive oxygen species and the precursor of hydroxyl radical.

persister: A member of a microbial population not killed by an antimicrobial but remains fully susceptible.

Petri dish: A covered glass or plastic dish in which the lid is slightly larger than the dish, allowing the lid to cover the dish in a way that enables oxygen access to the interior of the dish while generally keeping contaminating microbes outside.

pharmacodynamics: The effect of an antibiotic treatment on a microbe.

pharmacokinetics: The relationship between drug concentration and time following drug administration.

plaque: A hole in a lawn of bacteria created by infection of bacterial cells by a virus.

plasmid: An autonomous DNA molecule residing in a microbial cell.

polymerase chain reaction (PCR): A process by which a region of DNA is selectively amplified through repeated replication of the region by DNA polymerase.

post-antibiotic effect: Residual antibiotic action occurring after antibiotic concentration drops below toxic levels.

prevalence of resistance: The number of pathogen isolates having MIC greater than the interpretive breakpoint divided by the total number of isolates tested.

primer: A short nucleic acid used to initiate DNA replication at a specific location on a template nucleic acid.

probe: Generally a short nucleic acid used to detect the presence of a complementary nucleic acid in a patient sample.

promoter: A region of DNA, often upstream from a gene, at which RNA polymerase binds and begins transcription of the gene.

protein: Chain-like molecule composed of amino acids.

protein binding: Interaction of an antibiotic with serum or tissue; thought to reduce the effective concentration of antibiotics.

protein synthesis: The intracellular process of making new proteins; also called translation.

protozoan: A type of unicellular eukaryotic organism lacking a chloroplast or cell wall; examples are parasites that cause malaria and sleeping sickness.

quorum sensing: A process occurring in bacterial cultures in which small molecules released by one bacterium bind to another bacterial cell and cause a response in the second cell; the process is sensitive to cell density and therefore acts as a way for members of bacterial populations to sense when many members are present.

rational drug design: A process employed by the pharmaceutical industry in which the atomic structure of potential drugs and their biological targets are used to formulate new molecules likely to be highly active.

reassortment: The process occurring during influenza virus replication in which the eight viral RNA segments used to form mature virus particles are selected from a pool such that progeny viruses arising from an infection of a cell by two or more viruses contain a mixture of segments arising from the parental viruses.

recombination: Exchange of portions of DNA between two DNA molecules through a breakage and rejoining process.

relapse: Recurrence of disease following completion of treatment; generally associated with failure to remove the pathogen.

resistance mutations: Changes in the nucleotide sequence of DNA (RNA with same viruses) that enable an organism or virus to grow or at least survive treatment with an antibiotic at concentrations that inhibit growth or kill the parental organism or virus.

reverse transcription: The process of making a DNA molecule using RNA as a template.

ribosomal RNA (rRNA): A type of RNA that forms an integral part of ribosome structure.

ribosomes: Large intracellular structures composed of several ribosomal RNA molecules and many different ribosomal proteins; ribosomes serve as the intracellular sites where new proteins are made.

RNA: Ribonucleic acid; a long, thin chain-like molecule (nucleic acid) found in several varieties that have distinct functions.

RNA virus: A virus that uses RNA as its genetic material.

seasonal influenza: A respiratory disease caused by influenza virus that occurs each year during the winter and spring months.

serotype: A group of closely related microorganisms distinguished by a characteristic set of antigens, often thought of as a subspecies; also called a serovar.

site-specific recombination: DNA strand exchange taking place between segments possessing only a limited degree of nucleotide sequence homology; breaking and rejoining DNA at a specific location.

spore: A dormant body formed by certain bacteria, fungi, or other microbes that enables survival in adverse environmental conditions.

sputum: Material coughed from the respiratory tract, including saliva, mucus, and foreign material.

subpopulation: A minor subset of a heterogeneous population of microbes.

surveillance: The act of carefully watching; performing surveys to assess the prevalence and changes in prevalence of antibiotic resistance.

syncytium: A multinucleate mass of cytoplasm not separated into individual cells; created by infection with certain types of viruses.

template strand: The strand of DNA or RNA that serves as the source of nucleotide sequence information for formation of a second, complementary strand.

toxin-antitoxin module: A two-gene operon in which one gene encodes a poison and the other a protein that neutralizes the poison.

transduction: The process by which a virus moves genes from one cell to another.

transfer RNA (tRNA): A form of small RNA (generally shorter than 100 nucleotides) that serves as an adapter to order amino acids on ribosomes for incorporation into protein.

transformation: Change of form; with respect to bacteria, transformation refers to acquisition of a plasmid or other foreign DNA; with respect to animal cells, transformation refers to a normal cell becoming a tumor cell.

transposition: The process by which a transposon moves from one position in DNA to another. (Movement of many transposons is accompanied by making a copy of the transposon that moves.)

transposon: A segment of DNA capable of independently moving to a new position within the same or another DNA molecule.

vaccine: A biological preparation, often molecules characteristic of a pathogen or an attenuated pathogen, that enhances immunity to a specific disease by stimulating the immune system to recognize and destroy the pathogen when subsequently encountered.

viral receptors: Macromolecules, usually proteins or complex sugars, located on the surface of cells that enable attachment of virus particles.

virus: An infectious agent containing genetic material as either DNA or RNA surrounded by a protective protein coat; not capable of replication outside a living cell; generally submicroscopic.

yeast: A unicellular fungus.

Literature Cited

Chapter 1

1. Hawkings, N., Wood, F., Butler, C. "Public Attitudes Towards Bacterial Resistance: A Qualitative Study." *Journal of Antimicrobial Chemotherapy* 2007; 59:1155–1160.
2. Russell, S. "Family Still Battling Drug-Resistant Staph." *San Francisco Chronicle* 2008; January 20:A-10.
3. Hersh, A., Chambers, H., Maselli, J., Gonzales, R. "National Trends in Ambulatory Visits and Antibiotic Prescribing for Skin and Soft-Tissue Infections." *Archives of Internal Medicine* 2008; 168:1585–1591.
4. Klein, E., Smith, D., Laxminarayan, R. "Community-Associated Methicillin-Resistant Staphylococccus aureus in Outpatients, United States, 1999–2006." *Emerging Infectious Diseases* 2009; 15:1925–1930.
5. Klein, E., Smith, D., Laxminarayan, R. "Hospitalizations and Deaths Caused by Methicillin-Resistant *Staphylococcus aureus*, United States, 1999–2005." *Emerging Infectious Diseases* 2007; 13:1840–1846.
6. Zilberberg, M., Shorr, A., Kollef, M. "Growth and Geographic Variation in Hospitalizations with Resistant Infections, United States, 2000–2005." *Emerging Infectious Diseases* 2008; 14:1756–1758.
7. Nulens, E., Broex, E., Ament, A., Deurenberg, R., Smeets, E., Scheres, J., van Tiel, F., Gordts, B., Stobberingh, E. "Cost of the Methicillin-Resistant *Staphylococcus aureus* Search and Destroy Policy in a Dutch University Hospital." *Journal of Hospital Infection* 2008; 68:301–307.
8. Muder, R., Cunningham, C., McCray, E., Squier, C., Perreiah, P., Jain, R., Sinkowitz-Cochran, R., Jernigan, J. "Implementation of an Industrial Systems-Engineering Approach to Reduce the Incidence of Methicillin-Resistant *Staphylococcus aureus* Infection." *Infection Control and Hospital Epidemiology* 2008; 29:702–708.
9. Woolhouse, M.E. "Where Do Emerging Pathogens Come From?" *Microbe* 2006; 1:511–515.
10. Woolhouse, M.E., Gowtage-Sequeria, S. "Host Range and Emerging and Reemerging Pathogens." *Emerging Infectious Diseases* 2005; 11:1842–1847.
11. Turnbaugh, P., Ley, R., Mahowald, M., Magrini, V., Mardis, E., Gordon, J. "An Obesity-Associated Gut Microbiome with Increased Capacity for Energy Harvest." *Nature* 2006; 444:1027–1031.
12. Amaral, F., Sachs, D., Costa, V., Fagundes, C., Cisalpino, D., Cunha, T., Ferreira, S., Cunha, F., Silva, T., Nicoli, J., Vieira, L., Souza, D., Teixeira, M. "Commensal Microbiota Is Fundamental for the Development of Inflammatory Pain." *Proceedings of the National Academy of Sciences U.S.A.* 2008; 105:2193–2197.
13. Arias, C., Murray, B. "Antibiotic-Resistant Bugs in the 21st Century—a Clinical Super-Challenge." *New England Journal of Medicine* 2009; 360:439–443.
14. Shlaes, D., Projan, S., Edwards, J.E. "Antibiotic Discovery: State of the State." *ASM News* 2004; 70:275–281.
15. Mak J., Kim M., Pham J., Tapsall J., White P. "Antibiotic Resistance Determinants in Nosocomial Strains of Multidrug-Resistant *Acinetobacter baumannii*." *Journal of Antimicrobial Chemotherapy* 2009; 63:47–54.
16. Park Y., Peck K., Cheong H., Chung D.-R., Song J.-H., Ko K. "Extreme Drug Resistance in *Acinetobacter baumannii* Infections in Intensive Care Units, South Korea." *Emerging Infectious Diseases* 2009; 15:1325–1326.

17. Apisarnthanarak, A., Mundy, L. "Mortality Associated with Pandrug-Resistant *Acinetobacter baumannii* Infections in Thailand." *American Journal of Infection Control* 2009; 37:519–520.

18. Endimiani, A., Depasquale, J., Forero, S., Perez, F., Hujer, A., Roberts-Pollack, D., Fiorella, P., Pickens, N., Kitchel, B., Casiano-Colón, A., Tenover, F., Bonomo, R. "Emergence of *blaKPC*-Containing *Klebsiella pneumoniae* in a Long-Term Acute Care Hospital: A New Challenge to our Healthcare System." *Journal of Antimicrobial Chemotherapy* 2009; 64:1102–1110.

19. Hawkey, P., Jones, A. "The Changing Epidemiology of Resistance." *Journal of Antimicrobial Chemotherapy* 2009; 64:Suppl 1:i3–10.

20. Gandhi, N., Moll, A., Sturm, A., Pawinski, R., Govender, T., Lalloo, U., Zeller, K., Andrews, J., Friedland, G. "Extensively Drug-Resistant Tuberculosis as a Cause of Death in Patients Co-Infected with Tuberculosis and HIV in a Rural Area of South Africa." *The Lancet* 2006; 368:1575–1580.

21. Punga, V., Jakubowiak, E., Danilova, I., Somova, T., Volchenkov, G., Kazionnyy, B., Nemtsova, E., Kiryanova, E., Kourbatova, E. "Prevalence of Extensively Drug-Resistant Tuberculosis in Vladimir and Orel Regions, Russia." *International Journal of Tuberculosis and Lung Disease* 2009; 13:1309–1312.

22. Tapsall, J.W. "*Neisseria gonorrhoeae* and Emerging Resistance to Extended Spectrum Cephalosporins." *Current Opinion in Infectious Diseases* 2009; 223:87–91.

23. Matsumoto, T. "Trends of Sexually Transmitted Diseases and Antimicrobial Resistance in *Neisseria gonorrhoeae*." *International Journal of Antimicrobial Agents* 2008; 31S:S35–S39.

24. Barry, P., Klausner, J. "The Use of Cephalosporins for Gonorrhea: The Impending Problem of Resistance." *Expert Opinion in Pharmacotherapy* 2009; 10:555–577.

25. Zhu, M., Xie, Z., Zhang, L., Xia, S., Yang, W., Ran, L., Wang, Z. "Characterization of *Salmonella enterica* Serotype Typhimurium from Outpatients of 28 Hospitals in Henan Province in 2006." *Biomedical and Environmental Sciences* 2009; 22:136–140.

26. Krauland, M., Marsh, J., Paterson, D., Harrison, L. "Integron-Mediated Multidrug Resistance in a Global Collection of Nontyphoidal *Salmonella enterica* Isolates." *Emerging Infectious Diseases* 2009; 15:388–396.

27. Fluit, A., Wielders, C., Verhoef, J., Schmitz, F-J. "Epidemiology and Susceptibility of 3,051 *Staphylococcus aureus* Isolates from 25 University Hospitals Participating in the European SENTRY Study." *Journal of Clinical Microbiology* 2001; 39:3727–3732.

28. Woodford, N., Livermore, D. "Infections Caused by Gram-Positive Bacteria: A Review of the Global Challenge." *Journal of Infection* 2009; 59 Suppl. 1:S4–S16.

29. Pletz, M., Maus, U., Krug, N., Welte, T., Lode H. "Pneumococcal vaccines: Mechanism of Action, Impact on Epidemiology and Adaption of the Species." *International Journal of Antimicrobial Agents* 2008; 32:199–206.

30. Mooi, F., van Loo, H., van Gent, M., He, Q., Bart, M., Heuveman, K., de Greeff, S., Diavaopoulos, D., Teunis, P., Nagelkerke, N., Mertsola, J. "*Bordetella pertussis* Strains with Increased Toxin Production Associated with Pertussis Resurgence." *Emerging Infectious Diseases* 2009; 15:1206–1213.

31. Patel, S., Oshodi, A., Prasad, P., Delamora, P., Larson, E., Zaoutis, T., Paul, D., Saiman, L. "Antibiotic Use in Neonatal Intensive Care Units and Adherence with Centers for Disease Control and Prevention 12 Step Campaign to Prevent Antimicrobial Resistance." *Pediatric Infectious Disease Journal* 2009; 28:1047–1051.

32. Cespedes, A., Larson, E. "Knowledge, Attitudes, and Practices Regarding Antibiotic Use Among Latinos in the United States: Review and Recommendations." *American Journal of Infection Control* 2006; 34:495–502.

33. Larson, E., Dilone, J., Garcia, M., Smolowitz, J. "Factors Which Influence Latino Community Members to Self-Prescribe Antibiotics." *Nursing Research* 2006; 55:94–102.

Chapter 2

34. Tyagi, S., Kramer, F. "Molecular Beacons: Probes that Fluoresce upon Hybridization." *Nature Biotechnology* 1996; 14:303–308.

35. Varma-Basil, M., El-Hajj, H., Colangeli, R., Hazbón, M., Kumar, S., Bose, M., Bobadilla-del-Valle, M., García, L., Hernández, A., Kramer, F., Osornio, J., Ponce-de-León, A., Alland, D. "Rapid Detection of Rifampin Resistance in Mycobacterium tuberculosis Isolates from India and Mexico by a Molecular Beacon Assay." *Journal of Clinical Microbiology* 2004; 42:5512–5516.

36. Hillemann, D., Rüsch-Gerdes, S., Richter, E. "Feasibility of the Genotype MTBDRsl Assay for Fluoroquinolone, Amikacin-Capreomycin, and Ethambutol Resistance Testing of *Mycobacterium tuberculosis* Strains and Clinical Specimens." *Journal of Clinical Microbiology* 2009; 47:1767–1772.

37. Hillemann, D., Rüsch-Gerdes, S., Richter, E. "Evaluation of the Genotype MTBDRplus Assay for Rifampin and Isoniazid Susceptibility Testing of *Mycobacterium tuberculosis* Strains and Clinical Specimens." *Journal of Clinical Microbiology* 2007; 45:2635–2640.

38. Cohen, J. "The Duesberg Phenomenon." *Science* 1994; 266:1642–1643.

39. Abdool-Karim, S., Churchyard, G., Abdool-Karim, Q., Lawn, S. "HIV Infection and Tuberculosis in South Africa: An Urgent Need to Escalate the Public Health Response." *The Lancet* 2009; 374:921–933.

40. Duesberg, P. "HIV Is Not the Cause of AIDS." *Science* 1988; 241:514.

41. Duesberg, P., Koehnlein, C., Rasnick, D. "The Chemical Bases of the Various AIDS Epidemics: Recreational Drugs, Anti-Viral Chemotherapy and Malnutrition." *Journal of Biosciences* 2003; 28:383–412.

42. Cohen, J. "Could Drugs, Rather Than a Virus, Be the Cause of AIDS?" *Science* 1994; 266:1648–1649.

43. Cohen, J. "Fulfilling Koch's Postulates." *Science* 1994; 266.

44. Anonymous. "Blattner and Colleagues Respond to Duesberg." *Science* 1988; 241:514–517.

45. Blattner, W., Gallo, R., Temin, H. "HIV Causes AIDS." *Science* 1988; 241:515.

46. Cohen, J. "Duesberg and Critics Agree: Hemophilia Is the Best Test." *Science* 1994; 266:1645–1646.

47. Cohen, J. "The Epidemic in Thailand." *Science* 1994; 266:1647.

48. Moore, J. "A Duesberg, adieu!" *Nature* 1996; 380:293–294.

49. Chêne, G., Sterne, J., May, M., Costagliola, D., Ledergerber, B., Phillips, A., Dabis, F., Lundgren, J., Monforte, A., de-Wolf, F., Hogg, R., Reiss, P., Justice, A., Leport, C., Staszewski, S., Gill, J., Fatkenheuer, G., Egger, M. "Prognostic Importance of Initial Response in HIV-1 Infected Patients Starting Potent Antiretroviral Therapy: Analysis of Prospective Studies." *The Lancet* 2003; 362:679–686.

50. Espinosa, J., Herva, M., Andréoletti, O., Padilla, D., Lacroux, C., Cassard, H., Lantier, I., Castilla, J., Torres, J. "Transgenic Mice Expressing Porcine Prion Protein Resistant to Classical Scrapie but Susceptible to Sheep Bovine Spongiform Encephalopathy and Atypical Scrapie." *Emerging Infectious Diseases* 2009; 15:1214–1221.

51. Falkow, S. "Molecular Koch's Postulates Applied to Microbial Pathogenicity." Reviews of Infectious Diseases 1988; 10 Suppl 2:S274–S276.

52. Kiran, M., Giacomett, A., Cirioni, O., Balaban, N. "Suppression of Biofilm Related, Device-Associated Infections by staphylococcal Quorum Sensing Inhibitors." *International Journal of Artificial Organs* 2008; 31:761–770.

Chapter 3

53. Dixon, B. "Sulfa's True Significance." *Microbe* 2006; 1:500–501.

54. Alekshun, M., Levy, S. "Molecular Mechanisms of Antibacterial Multidrug Resistance." *Cell* 2007; 128:1037–1050.

55. Shah, N., Alker, A., Sem, R., Susanti, A., Muth, S., Maguire, J., Duong, S., Ariey, F., Meshnick, S., Wongsrichanalai, C. "Molecular Surveillance for Multidrug-Resistant *Plasmodium falciparum,* Cambodia." *Emerging Infectious Diseases* 2008; 14:1637–1640.

56. Farmer, T., Gilbart, J., Elson, S. "Biochemical Basis of Mupirocin Resistance in Strains of *Staphylococcus aureus*." *Journal of Antimicrobial Chemotherapy* 1992; 30:587–596.

57. O'Neill, A., Chopra, I. "Molecular Basis of fusB-Mediated Resistance to Fusidic Acid in *Staphylococcus aureus*." *Molecular Microbiology* 2006; 59:664–676.

58. Sidhu, A., Verdier-Pinard, D., Fidock, D.A. "Chloroquine Resistance in *Plasmodium falciparum* Malaria Parasites Conferred by pfcrt mutations." *Science* 2002; 298:210–213.

59. Ernst, C., Staubitz, P., Mishra, N., Yang, S., Hornig, G., Kalbacher, H., Bayer, A., Kraus, D., Peschel, A. "The Bacterial Defensin Resistance Protein MprF Consists of Separable Domains for Lipid Lysinylation and Antimicrobial Peptide Repulsion." *PLoS Pathogens* 2009; 5:e1000660.

60. Jhingran, A., Chawla, B., Saxena, S., Barrett, M., Madhubala, R. "Paromomycin: Uptake and Resistance in *Leishmania donovani*." *Molecular and Biochemical Parasitology* 2009; 164:111–117.

61. Duan, R., de Vries, R., van Dun, J., van Loenen, F., Osterhaus, A., Remeijer, L., Verjans, G. "Acyclovir Susceptibility and Genetic Characteristics of Sequential Herpes Simplex Virus Type 1 Corneal Isolates from Patients with Recurrent Herpetic Keratitis." *Journal of Infectious Diseases* 2009; 200:1402–1414.

62. Geerts, S., Gryseels, B. "Drug Resistance in Human Helminths: Current Situation and Lessons from Livestock." *Clinical Microbiology Reviews* 2000; 13:207–222.

63. Favre, B., Ghannoum, M., Ryder, N. "Biochemical Characterization of Terbinafine-Resistant Trichophyton rubrum Isolates." *Medical Mycology* 2004; 42:525–529.

64. Nielsen-Kahn, J., Garcia-Effron, G., Hsu, M.-J., Park, S., Marr, K.A., Perlin, D.S. "Acquired Echinocandin Resistance in a *Candida krusei* Isolate Due to Modification of Glucan Synthase." *Antimicrobial Agents and Chemotherapy* 2007; 51:1876–1878.

65. Papon, N., Noël T., Florent M., Gibot-Leclerc, S., Jean, D., Chastin, C., Villard, J., Chapeland-Leclerc, F. "Molecular Mechanism of Flucytosine Resistance in Candida lusitaniae: Contribution of the FCY2, FCY1, and FUR1 Genes to 5-fluorouracil and Fluconazole Cross-Resistance." *Antimicrobial Agents and Chemotherapy* 2007; 51:369–371.

66. Paredes, R., Clotet, B. "Clinical Management of HIV-1 Resistance." *Antiviral Research* 2010; 85:245-265.

67. Perichon, B., Courvalin, P. "VanA-type Vancomycin-Resistant *Staphylococcus aureus*." *Antimicrobial Agents and Chemotherapy* 2009; 53:4580–4587.

68. Finks, J., Wells, E., Dyke, T., Husain, N., Plizga, L., Heddurshetti, R., Wilkins, M., Rudrik, J., Hageman, J., Patel, J., Miller, C. "Vancomycin-Resistant *Staphylococcus aureus*, Michigan, USA, 2007." *Emerging Infectious Diseases* 2009; 15:943–945.

69. Anonymous. Sulfonamide (Medicine): Wikipedia, 2008.

70. Anonymous. U.S. Food and Drug Administration: Wikipedia, 2008.

71. Dwyer, D., Kohanski, M., Hayete, B., Collins, J. "Gyrase Inhibitors Induce an Oxidative Damage Cellular Death Pathway in Escherichia coli." *Molecular and Systems Biology* 2007; 3:91. Epub.

72. Kohanski, M., Dwyer, D., Hayete, B., Lawrence, C., Collins, J. "A Common Mechanism of Cellular Death Induced by Bactericidal Antibiotics." *Cell* 2007; 130:797–810.

73. Wang, X., Zhao, X. "Contribution of Oxidative Damage to Antimicrobial Lethality." *Antimicrobial Agents and Chemotherapy* 2009; 53:1395–1402.

74. MacKenzie, W., Hoxie, N., Proctor, M., Gradus, M., Blair, K., Peterson, D., Kazmierczak, J., Addiss, D., Fox, K., Rose, J., Davis, J. "A Massive Outbreak in Milwaukee of Cryptosporidium Infection Transmitted Through the Public Water Supply." *New England Journal of Medicine* 1994; 331:161–167.

75. Acar, J., Goldstein, F. "Trends in Bacterial Resistance to Fluoroquinolones." *Clinical Infectious Diseases* 1997; 24:S67–S73.

76. Chen, D.K., McGeer, A., deAzavedo J.C., Low, D.E. "Decreased Susceptibility of *Streptococcus pneumoniae* to Fluoroquinolones in Canada." *New England Journal of Medicine* 1999; 341:233–239.

77. Park-Wyllie, L., Juurlink, D., Kopp, A., Shah, B., Stukel, T., Stumpo, C., Dresser, L., Low, D., Mamdani, M. "Outpatient Gatifloxacin Therapy and Dysglycemia in Older Adults." *New England Journal of Medicine* 2006; 354:1352–1361.

78. von Gottberg, A., Klugman, K., Cohen, C., Wolter, N., de Gouveia, L., du Plessis, M., Mpembe, R., Quan, V., Whitelaw, A., Hoffmann, R., Govender, N., Meiring, S., Smith, A., Schrag, S. "Emergence of Levofloxacin-Non-Susceptible *Streptococcus pneumoniae* and Treatment for Multidrug-Resistant Tuberculosis in Children in South Africa: A Cohort Observational Surveillance Study." *The Lancet* 2008; 371:1108–1113.

79. Perlman, D., ElSadr, W., Heifets, L., Nelson, E., Matts, J., Chirgwin, K., Salomon, N., Telzak, E., Klein, O., Kreiswirth, B., Musser, J., Hafner, R. "Susceptibility to Levofloxacin of *Mycobacterium tuberculosis* Isolates from Patients with HIV-Related Tuberculosis and Characterization of a Strain with Levofloxacin Monoresistance." *AIDS* 1997; 11:1473–1478.

80. Drlica, K., Zhao, X., Kreiswirth, B. "Minimizing Moxifloxacin Resistance with Tuberculosis." *Lancet Infectious Diseases* 2008; 8:273–275.

81. German, N., Malik, M., Rosen, J., Drlica, K., Kerns, R. "Use of Gyrase Resistance Mutants to Guide Selection of 8-Methoxy-Quinazoline-2,4-Diones." *Antimicrobial Agents and Chemotherapy* 2008; 52:3915–3921.

82. Malik, M., Hoatam, G., Chavda, K., Kerns, R., Drlica. K. "Novel Approach for Comparing Quinolones for Emergence of Resistant Mutants During Quinolone Exposure." *Antimicrobial Agents and Chemotherapy* 2010; 54:149–156.

Chapter 4

83. Ley, R., Hamady, M., Lozupone, C., Turnbaugh, P., Ramey, R., Bircher, J., Schlegel, M., Tucker, T., Schrenzel, M., Knight, R., Gordon, J. "Evolution of Mammals and Their Gut Microbes." *Science* 2008; 320:1647–1651.

84. Kahlmeter, G., Brown, D. "Harmonization of Antimicrobial Breakpoints in Europe—Can It Be Achieved?" *Clinical Microbiology Newsletter* 2004; 26:187–192.

85. David, M., Crawford, S., Boyle-Vavra, S., Hostetler, M., Kim, D., Daum, R. "Contrasting Pediatric and Adult Methicillin-Resistant Isolates." *Emerging Infectious Diseases* 2006; 12:631–637.

86. Shin, S., Yagui, M., Ascencios, L., Yale, G., Suarez, C., Quispe, N., Bonilla, C., Blaya, J., Taylor, A., Contreras, C., Cegielski, P., "Scale-Up of Multidrug-Resistant Tuberculosis Laboratory Services, Peru." *Emerging Infectious Diseases* 2008; 14:701–708.

87. Bonilla, C., Crossa, A., Jave, H., Mitnick, C., Jamanca, R., Herrera, C., Asencios, L., Mendoza, A., Bayona, J., Zignol, M., Jaramillo, E. "Management of Extensively Drug-Resistant Tuberculosis in Peru: Cure Is Possible." *PLos ONE* 2008; 3:e2957.

88. Gumbo, T., Louie, A., Deziel, M., Parsons, L., Salfinger, M., Drusano, G. "Selection of a Moxifloxacin Dose That Suppresses Drug Resistance in *Mycobacterium tuberculosis* by Use of an

In Vitro Pharmacodynamic Infection Model and Mathematical Modeling." *Journal of Infectious Diseases* 2004; 190:1642–1651.

89. Craig, W.A. "Pharmacodynamics of Antimicrobials: General Concepts and Applications." In: Nightingale, C., Murakawa, T., Ambrose, P, eds. *Antimicrobial Pharmacodynamics in Theory and Clinical Practice* New York: Marcel Dekker, 2002:1–22.

90. Craig, W.A., Andes, D.R. "Correlation of the Magnitude of the AUC_{24}/MIC for 6 Fluoroquinolones Against *Streptococcus pneumoniae* (SP) with Survival and Bactericidal Activity in an Animal Model," 40th ICAAC, 2000.

91. Menson, E., Walker, A., Sharland, M., Wells, C., Tudor-Williams, G., Riordan, F., Lyall, E., Gibb, D., and The Collaborative-HIV-Paediatric-Study-Steering-Committee. "Underdosing of Antiretrovirals in UK and Irish Children with HIV as an Example of Problems in Prescribing Medicines to Children, 1997–2005: Cohort Study." *British Medical Journal* 2006; 332:1183–1187.

92. Lehrnbecher, T., Kaiser, J., Varwig, D., Ritter, J., Groll, A., Creutzig, U., Klingebiel, T., Schwabe, D. "Antifungal Usage in Children Undergoing Intensive Treatment for Acute Myeloid Leukemia: Analysis of the Multicenter Clinical Trial AML-BFM 93." *European Journal of Clinical Microbiology and Infectious Diseases* 2007; 26:735–738.

93. Paci, P., Carello, R., Bernaschi, M., D'Offizi, G., Castiglione, F. "Immune Control of HIV-1 Infection After Therapy Interruption: Immediate versus Deferred Antiretroviral Therapy." *BMC Infectious Diseases* 2009; 9:172.

94. World-Health-Organization. "Anti-Tuberculosis Drug Resistance in the World." Report No. 4. Geneva, 2008:WHO/HTB/TB/2008.394.

95. Bergval, I., Schuitema, A., Klatser, P., Anthony, R. "Resistant Mutants of *Mycobacterium tuberculosis* Selected In Vitro Do Not Reflect the In Vivo Mechanism of Isoniazid Resistance." *Journal of Antimicrobial Chemotherapy* 2009; 64:515–523.

96. Stronati, M., Borghesi, A., Decembrino, L., Bollani, L. "Antibiotics in Neonatal Intensive Care Units (NICUs)." *Journal of Antimicrobial Chemotherapy* 2007; Suppl 2:52–55.

97. Grigoryan, L., Haaijer-Rysjamp, F., Burgerhof, J., Mechtler, R., Deschepper, R., Tambic-Andrasevic, A,, Andrajati, R., Monnet, D., Cunney, R., DiMatteo, A., Edelsein, H., Valinteliene, R., Alkerwi, A., Scicluna, E., Grzesiowski, P., Bara, A., Tesar, T., Cizman, M., Campos, J., Lundborg, C., Birkin, J. "Self-Medication with Antimicrobial Drugs in Europe." *Emerging Infectious Diseases* 2006; 12:452–459.

98. McNulty, C., Boyle, P., Nichols, T., Clappison, D., Davey, P. "Antimicrobial Drugs in the Home, United Kingdom." *Emerging Infectious Diseases* 2006; 12:1523–1526.

99. Mainous, A., Cheng, A., Garr R., Tilley, B., Everett, C., McKee M. "Nonprescribed Antimicrobial Drugs in Latino Community, South Carolina." *Emerging Infectious Diseases* 2005; 11:883–888.

100. Cespedes, A., Larson, E. "Knowledge, Attitudes, and Practices Regarding Antibiotic Use Among Latinos in the United States: Review and Recommendations." *American Journal of Infection Control* 2006; 34:495–502.

101. Larson, E., Dilone, J., Garcia, M., Smolowitz, J. "Factors Which Influence Latino Community Members to self-prescribe Antibiotics." *Nursing Research* 2006; 55:94–102.

102. Andes, D., Craig, W. "Pharmacodynamics of the New Fluoroquinolone Gatifloxacin in Murine Thigh and Lung Infection Models." *Antimicrobial Agents and Chemotherapy* 2002; 46:1665–1670.

Chapter 5

103. Sieradzki, K., Leski, T., Dick, J., Borio, L., Tomasz, A. "Evolution of a Vancomycin-Intermediate *Staphylococcus aureus* Strain In Vivo: Multiple Changes in the Antibiotic Resistance Phenotypes of a Single Lineage of Methicillin-Resistant *S. aureus* Under the Impact of Antibiotics Administered for Chemotherapy." *Journal of Clinical Microbiology* 2003; 41:1687–1693.

104. Davidson, R., Cavalcanti, R., Brunton, J., Bast, D., deAzavedo, J., Kibsey, P., Fleming, C., Low, D. "Resistance to Levofloxacin and Failure of Treatment of Pneumococcal Pneumonia." *New England Journal of Medicine* 2002; 346:747–750.

105. Cui, J., Liu, Y., Wang, R., Tong, W., Drlica, K., Zhao, X. "The Mutant Selection Window Demonstrated in Rabbits Infected with *Staphylococcus aureus*." *Journal of Infectious Diseases* 2006; 194:1601–1608.

106. Bridges, B. "Hypermutation in Bacteria and Other Cellular Systems." *Philosophical Transactions of the Royal Society London B Biological Science* 2001; 356:29–39.

107. Urban, C., Rahman, N., Zhao, X., Mariano, N., Segal-Maurer, S., Drlica, K., Rahal, J. "Fluoroquinolone-Resistant *Streptococcus pneumoniae* Associated with Levofloxacin Therapy." *Journal of Infectious Diseases* 2001; 184:794–798.

108. Liu, Y., Cui, J., Wang, R., Wang, X., Drlica, K., Zhao, X. "Selection of Rifampicin–Resistant *Staphylococcus aureus* During tuberculosis Therapy: Concurrent Bacterial Eradication and Acquisition of Resistance." *Journal of Antimicrobial Chemotherapy* 2005; 56:1172–1175.

109. Zhao, X., Drlica, K. "Restricting the Selection of Antibiotic-Resistant Mutants: a General Strategy Derived from Fluoroquinolone Studies." *Clinical Infectious Diseases* 2001; 33 (Suppl 3):S147–S156.

110. Drlica, K., Zhao, X. "Mutant Selection Window Hypothesis Updated." *Clinical Infectious Diseases* 2007; 44:681–688.

111. Zhou, J., Dong, Y., Zhao, X., Lee, S., Amin, A., Ramaswamy, S., Domagala, J., Musser, J., Drlica, K. "Selection of Antibiotic Resistant Bacterial Mutants: Allelic Diversity Among Fluoroquinolone-Resistant Mutations." *Journal of Infectious Diseases* 2000; 182:517–525.

112. Blondeau, J., Zhao, X., Hansen, G., Drlica, K. "Mutant Prevention Concentrations (MPC) for Fluoroquinolones with Clinical Isolates of *Streptococcus pneumoniae*." *Antimicrobial Agents and Chemotherapy* 2001; 45:433–438.

113. Quinn, B., Hussain, S., Malik, M., Drlica, K., Zhao, X. "Daptomycin Inoculum Effects and Mutant Prevention Concentration with *Staphylococcus aureus*." *Journal of Antimicrobial Chemotherapy* 2007; 60:1380–1383.

114. Firsov, A., Vostrov, S., Lubenko, I., Drlica, K., Portnoy, Y., Zinner, S. "In Vitro Pharmacodynamic Evaluation of the Mutant Selection Window Hypothesis: Four Fluoroquinolones Against *Staphylococcus aureus*." *Antimicrobial Agents and Chemotherapy* 2003; 47:1604–1613.

115. Zinner, S., Lubenko, I., Gilbert, D., Simmons, K., Zhao, X., Drlica, K., Firsov, A. "Emergence of Resistant *Streptococcus pneumoniae* in an In Vitro Dynamic Model That Simulates Moxifloxacin Concentration In and Out of the Mutant Selection Window: Related Changes in Susceptibility, Resistance Frequency, and Bacterial Killing." *Journal of Antimicrobial Chemotherapy* 2003; 52:616–622.

116. Cirz, R., Chin, J., Andes, D., Crecy-Lagard, V., Craig, W., Romesberg, F.E. "Inhibition of Mutation and Combating the Evolution of Antibiotic Resistance." *Plos Biology* 2005; 3:1024–1033.

117. Cirz, R., Romesberg, F. "Induction and Inhibition of Ciprofloxacin Resistance-Conferring Mutations in Hypermutator Bacteria." *Antimicrobial Agents and Chemotherapy* 2006; 50:220–225.

118. Poole, K. "Efflux-Mediated Antimicrobial Resistance." *Journal of Antimicrobial Chemotherapy* 2005; 56:20–51.

119. Fraud, S., Campigotto, A., Chen, Z., Poole, K. "The MexCD-OprJ Multidrug Efflux System of *Pseudomonas aeruginosa*: Involvement in Chlorhexidine Resistance and Induction by Membrane Damaging Agents Dependent Upon the AlgU Stress-Response Sigma Factor." *Antimicrobial Agents and Chemotherapy* 2008; 52:4478–4482.

120. Jeannot, K., Elsen, S., Köhler, T., Attree, I., van Delden, C., Plésiat, P. "Resistance and Virulence of *Pseudomonas aeruginosa* Clinical Strains Overproducing the MexCD-OprJ Efflux Pump." *Antimicrobial Agents and Chemotherapy* 2008; 52:2455–2462.

121. Rice, L. "Evolution and Clinical Importance of Extended-Spectrum β-Lactamases." *Chest* 2001; 119:391–396.

122. Datta, N., Kontomichalou, P. "Penicillinase Synthesis Controlled by Infectious R Factors in Enterobacteriaceae." *Nature* 1965; 208:239–241.

123. Saudagar, P., Survase, S., Singhal, R. "Clavulanic Acid: a Review." *Biotechnology Advances* 2008; 26:335–351.

124. Song, S., Berg, O., Roth, J., Andersson, D. "Contribution of Gene Amplification to Evolution of increased Antibiotic Resistance in *Salmonella typhimurium*." *Genetics* 2009; 182:1183–1195.

125. Hegde, S., Vetting, M., Roderick, S., Mitchenall, L., Maxwell, A., Takiff, H., Blanchard, J. "A Fluoroquinolone Resistance Protein from *Mycobacterium tuberculosis* That Mimics DNA." *Science* 2005; 308:1480–1483.

126. Miller, J. "Spontaneous Mutators in Bacteria: Insights into Pathways of Mutagenesis and Repair." *Annual Review of Microbiology* 1996; 50:625–643.

127. Oliver, A., Levin, B., Juan, C., Baquero, F., Blazquez, J. ""Hypermutation and the Preexistence of Antibiotic-Resistant *Pseudomonas aeruginosa* Mutants: Implications for Susceptibility Testing and Treatment of Chronic Infections. *Antimicrobial Agents and Chemotherapy* 2004; 48:4226–4233.

128. Rhee, K., Sethupathi, P., Driks, A., Lanning, D., Knight, K. "Role of Commensal Bacteria in Development of Gut-Associated Lymphoid Tissues and Preimmune Antibody Repertoire." *Journal of Immunology* 2004; 172:1118–1124.

129. Poole, K. "Efflux-Mediated Antimicrobial Resistance." *Journal of Antimicrobial Chemotherapy* 2005; 56:20–51.

130. Domingo, A., Holland, J. "RNA Virus Mutations and Fitness for Survival." *Annual Review of Microbiology* 1997; 51:151–178.

131. Metzner, K., Giulieri, S., Knoepfel, S., Rauch, P., Burgisser, P., Yerly, S., Günthard, H., Cavassini, M. "Minority Quasispecies of Drug-Resistant HIV-1 That Lead to Early Therapy Failure in Treatment-Naive and -Adherent Patients." *Journal of Clinical Microbiology* 2009; 48:239–247.

132. Le, T., Chiarella, J., Simen, B., Hanczaruk, B., Egholm, M., Landry,M., Dieckhaus, K., Rosen, M., Kozal, M. "Low-Abundance HIV Drug-Resistant Viral Variants in Treatment-Experienced Persons Correlate with Historical Antiretroviral Use." *PLoS ONE* 2009; 4:e6078.

133. Little, S., Frost, S., Wong, J., Smith, D., Pond, S., Ignacio, C., Parkin, N., Petropoulos, C., Richman, D. "Persistence of Transmitted Drug Resistance Among Subjects with Primary Human Immunodeficiency Virus Infection." *Journal of Virology* 2008; 82:5510–5518.

134. Ross, L., Lim, M., Liao, Q., Wine, B., Rodriguez, A., Weinberg, W., Shaefer, M. "Prevalence of Antiretroviral Drug Resistance and Resistance-Associated Mutations in Antiretroviral Therapy-Naïve HIV Infected Individuals from 40 United States Cities." *HIV Clinical Trials* 2007; 8:1–8.

135. Aarestrup, F.M. "Monitoring of Antimicrobial Resistance Among Food Animals: Principles and Limitations." *Journal of Veterinary Medicine* 2004; B51:380–388.

136. Songer, J., Trinh, H., Killgore, G., Thompson, A., McDonald, L., Limbago, B. "*Clostridium difficile* in Retail Meat Products, USA, 2007." *Emerging Infectious Diseases* 2009; 15:819–821.

137. Rodriguez-Palacios, A., Reid-Smith, R., Staempfli, H., Daignault, D., Janecko, N., Avery, B., Martin, H., Thomspon, A., McDonald, L., Limbago, B., Weese, J. "Possible Seasonality of *Clostridium difficile* in Retail Meat, Canada." *Emerging Infectious Diseases* 2009; 15:802–805.

138. Bakri, M., Brown, D., Butcher, J., Sutherland, A. "*Clostridium difficile* in Ready-to-Eat Salads, Scotland." *Emerging Infectious Diseases* 2009; 15:817–818.

139. Zilberberg, M., Shorr, A., Kollef, M. "Increase in Adult *Clostridium difficile*-Related Hospitalizations and Case-Fatality Rate, United States, 2000–2005." *Emerging Infectious Diseases* 2008; 14:929–931.

140. Jagal, J., Naumova E. "*Clostridium difficile*-Associated Disease in the Elderly, United States." *Emerging Infectious Diseases* 2009; 15:343–344.

141. McDonald, L., Owings, M., Jernigan, D. "*Clostridium difficile* Infection in Patients Discharged from US Short-Stay Hospitals, 1996–2003." *Emerging Infectious Diseases* 2006; 12:409–415.

142. Bartlett, J., Gerding, D. "Clinical Recognition and Diagnosis of *Clostridium difficle* Infection." *Clinical Infectious Disease* 2008; 46:S12–S18.

143. O'Brien, J., Lahue, B., Caro, J., Davidson, D. "The Emerging Infectious Challenge of *Clostridium difficile*-Associated Disease in Massachusetts Hospitals: Clinical and Economic Consequences." *Infection Control and Hospital Epidemiology* 2007; 28:1219–1227.

144. Davidson, R., Davis, I., Willey, B., Rizg, K., Bolotin, S., Porter, V., Polsky, J., Daneman, N., McGeer, A., Yang, P., Scolnik, D., Rowsell, R., Imas, O., Silverman, M. "Antimalarial Therapy Selection for Quinolone Resistance Among *Escherichia coli* in the Absence of Quinolone Exposure in Tropical South America." *PLoS ONE* 2008; 3:e2727.

Chapter 6

145. Chhibber, S., Kaur, S., Kumari, S. "Therapeutic Potential of Bacteriophage in Treating *Klebsiella pneumoniae* B5055-Mediated Lobar Pneumonia in Mice." *Journal of Medical Microbiology* 2008; 57:1508–1513.

146. Garau, J., Xercavins, M., Rodriguez-Carballerira, M., Gomez-Vera, J., Coll, I., Vidal, D., Llovet, T., Ruiz-Bremon, A. "Emergence and Dissemination of Quinolone-Resistant *Escherichia coli* in the Community." *Antimicrobial Agents and Chemotherapy* 1999; 43:2736–2741.

147. Fantin, B., Duval, X., Massias, L., Alavoine, L., Chau, F., Retout, S., Andremont, A., Mentré, F. "Ciprofloxacin Dosage and Emergence of Resistance in Human Commensal Bacteria." *Journal of Infectious Diseases* 2009; 200:390–398.

148. Bartoloni, A., Pallecchi, L., Fiorelli, C., DiMaggio, T., Fernandez, C., Villagran, A., Mantella, A., Bartalesi, F., Strohmeyer, M., Bechini, A., Gamboa, H., Rodriguez, H., Kristiansson, C., Kronvall, G., Gotuzzo, E., Paradisi, F., Rossolini, G. "Increasing Resistance in Commensal *Escherichia coli*, Bolivia and Peru." *Emerging Infectious Diseases* 2008; 14:338–340.

149. Mazel, D. "Integrons: Agents of Bacterial Evolution." *Nature Reviews in Microbiology* 2006; 4:608–620.

150. Márquez, C., Labbate, M., Raymondo, C., Fernández, J., Gestal, A., Holley, M., Borthagaray, G., Stokes, H. "Urinary Tract Infections in a South American Population: Dynamic Spread of Class 1 Integrons and Multidrug Resistance by Homologous and Site-specific Recombination." *Journal of Clinical Microbiology* 2008; 46:3417–3425.

151. Krauland, M., Marsh, J., Paterson, D., Harrison, L. "Integron-Mediated Multidrug Resistance in a Global Collection of Nontyphoidal *Salmonella enterica* Isolates." *Emerging Infectious Diseases* 2009; 15:388–396.

152. Coburn, P., Baghdayan, A., Dolan, G., Shankar, N. "Horizontal Transfer of Virulence Genes Encoded on the *Enterococcus faecalis* Pathogenicity Island." *Molecular Microbiology* 2006; 63:530–544.

153. Lujan, S.A G.L, Ragonese, H., Matson, S.W., Redinbo, M.R. "Disrupting Antibiotic Resistance propagation by Inhibiting the Conjugative DNA Relaxase." *Proceedings of the National Academy of Sciences U.S.A.* 2007; 104:12282–12287.

154. Wang, X., Zhao, X., Malik, M., Drlica, K. "Contribution of Reactive Oxygen Species to Pathways of Quinolone-Mediated Bacterial Cell Death." *Journal of Antimicrobial Chemotherapy* 2010; 65:520-524.

Chapter 7

155. Reichman, L.B., Tanne, J. *Timebomb: The Global Epidemic of Multi-Drug-Resistant Tuberculosis.* New York: McGraw-Hill, 2002:240.

156. Bearden, D., Allen, G. "Impact of Antimicrobial Control Programs on Patient Outcomes." *Disease Management and Health Outcomes* 2003; 11:723–736.

157. Escombe, A., Moore, D., Gilman, R., Pan, W., Navincopa, M., Ticona, E., Martínez, C., Caviedes, L., Sheen, P., Gonzalez, A., Noakes, C., Friedland, J., Evans, C. "The Infectiousness of Tuberculosis Patients Coinfected with HIV." *Plos Medicine* 2008; 5:e188.

158. McAdam, J., Bucher, S., Brickner, P., Vincent, R., Lascher, S. "Latent Tuberculosis and Active Tuberculosis Disease Rates Among the Homeless, New York, New York, USA, 1992–2006." *Emerging Infectious Diseases* 2009; 15:1109–1111.

159. Xu, C., Kreiswirth, B.N., Sreevatsan, S., Musser, J.M., Drlica, K. "Fluoroquinolone Resistance Associated with Specific Gyrase Mutations in Clinical isolates of Multidrug Resistant *Mycobacterium tuberculosis*." *Journal of Infectious Diseases* 1996; 174:1127–1130.

160. Frieden, T., Fujiwara, P., Washko, R., Hamburg, M. "Tuberculosis in New York City—Turning the Tide." *New England Journal of Medicine* 1995; 333:229–233.

161. White, V., Moore-Gillon, J. "Resource Implications of Patients with Multidrug Resistant Tuberculosis." *Thorax* 2000; 55:962–963.

162. Dewan, R., Sosnovskaja, A., Thomsen, V., Cicenaite, J., Laseson, K., Johansen, I., Davidaviciene, E., Wells, C., "High Prevalence of Drug-Resistant Tuberculosis, Republic of Lithuania, 2002." *International Journal of Tuberculosis and Lung Disease* 2005; 9:170–174.

163. Badiaga, S., Raoult, D., Brouqui, P. "Preventing and Controlling Emerging and Reemerging transmissible Diseases in the Homeless." *Emerging Infectious Diseases* 2008; 14:1353–1359.

164. Tellier, R. "Review of Aerosol Transmission of Influenza A Virus." *Emerging Infectious Diseases* 2006; 12:1657–1662.

165. Suzita, R., Abdulamir, A., Bakar, F., Son R. "A Mini Review: Cholera Outbreak via Shellfish." *American Journal of Infectious Disease* 2009; 5:40–47.

166. Sagel, U., Schulte, B., Heeg, P., Borgmann, S. "Vancomycin-Resistant Enterococci Outbreak, Germany, and Calculation of Outbreak Start." *Emerging Infectious Diseases* 2008; 14:317–319.

167. Whittington, A., Whitlow, G., Hewson, D., Thomas, C., Brett, S. "Bacterial Contamination of Stethoscopes on the Intensive Care Unit." *Anaesthesia* 2009; 64:620–624.

168. Zinderman, C., Conner, B., Malakooti, M., LaMar, J., Armstrong, A., Bohnker, B. "Community-Acquired Methicillin-Resistant *Staphylococcus aureus* Among Military Recruits." *Emerging Infectious Diseases* 2004; 10:941–944.

169. Kazakova, S., Hageman, J., Matava, M., Srinivasan, A., Phelan, L., Garfinkel, B., Boo, T., McAllister, S., Anderson, J., Jensen, B., Dodson, D., Lonsway, D., McDougal, L., Arduino, M., Fraser, V., Killgore, G., Tenover, F., Cody, S., Jernigan, D. "A Clone of Methicillin-Resistant *Staphylococcus aureus* Among Professional Football Players." *New England Journal of Medicine* 2005; 352:468–475.

170. Huijsdens, X., van Lier, A., van Kregten, E., Verhoef, L., van Santen-Verheuvel, M., Spalburg, E., Wannet, W. "Methicillin-Resistant *Staphylococcus aureus* in Dutch Soccer Team." *Emerging Infectious Diseases* 2006; 12:1584–1586.

171. Garza, D., Sungar, G., Johnston, T., Rolston, B., Ferguson, J., Matheson, G. "Ineffectiveness of Surveillance to Control Community-Acquired Methicillin-Resistant *Staphylococcus aureus* in a Professional Football Team." *Clinical Journal of Sports Medicine* 2009; 19:498–501.

172. Freedman, D.O. "Clinical Practice: Malaria Prevention in Short-Term Travelers." *New England Journal of Medicine* 2008; 359:603–612.

173. Bechah, Y., Capo, C., Mege, J., Raoult, D. "Epidemic Typhus." *Lancet Infect Diseases* 2008; 8:417–426.

174. Reisen, W., Takahashi, R., Carroll, B., Quiring, R. "Delinquent Mortgages, Neglected Swimming Pools, and West Nile Virus, California." *Emerging Infectious Diseases* 2008; 14:1747–1749.

175. Goodman, R., Buehler, J. "Delinquent Mortgages, Neglected Swimming Pools, and West Nile Virus, California." *Emerging Infectious Diseases* 2009; 15:508.

176. Reisen, W., Takahashi, R., Carroll, B., Quiring, R. "Delinquent Mortgages, Neglected Swimming Pools, and West Nile Virus, California." *Emerging Infectious Diseases* 2009; 15:508–509.

177. Groopman, J. "Superbugs." *The New Yorker*, August 11, 2008:46–55.

178. Quale, J. "Global Spread of Carbapenemase-Producing *Klebsiella pneumoniae*." *Microbe* 2008; 3:516–520.

179. Clock, S., Cohen, B., Behta, M., Ross, B., Larson, E. "Contact Precautions for Multidrug-Resistant Organisms: Current Recommendations and Actual Practice." *American Journal of Infection Control* 2010; 38:105–111.

180. Larson, E. "Hygiene of the Skin: When Is Clean Too Clean." *Emerging Infectious Diseases* 2001; 7:225–230.

Chapter 8

181. Hall, G., Yohannes, K., Raupach, J., Becker, N., Kirk, M. "Estimating Community Incidence of *Salmonella*, *Campylobacter*, and Shiga Toxin-Producing *Escherichia coli* Infections, Australia." *Emerging Infectious Diseases* 2008; 14:1601–1609.

182. Masterton, R. "The Importance and Future of Antimicrobial Surveillance Studies." *Clinical Infectious Diseases* 2008; 47:S21–S31.

183. Tyagi, S., Bratu, D., Kramer, F. "Multicolor Molecular Beacons for Allele Discrimination." *Nature Biotechnology* 1998; 16:49–53.

184. El-Hajj, H., Marras, S., Tyagi, S., Shashkina, E., Kamboj, M., Kiehn, T., Glickman, M., Kramerj, F., Alland, A. "Use of Sloppy Molecular Beacon Probes for Identification of Mycobacterial Species." *Journal of Clinical Microbiology* 2009; 47:1190–1198.

185. Karlowsky, J., Sahm, D. "Antibiotic Resistance—Is Resistance Detected by Surveillance Relevant to Predicting Resistance in the Clinical Setting?" *Current Opinion in Pharmacology* 2002; 2:487–492.

186. Morris, A., Masterton, R. "Antibiotic Resistance Surveillance: Action for International Studies." *Journal of Antimicrobial Chemotherapy* 2002; 49:7–10.

187. Nelson, J., Chiller, T., Powers, J., Angulo, F. "Fluoroquinolone-Resistant *Campylobacter* Species and the Withdrawal of Fluoroquinolones from Use in Poultry: A Public Health Success Story." *Clinical Infectious Diseases* 2007; 44:977–980.

188. Willems, R., Top, J., vanSanten, M., Robinson, D., Coque, T., Baquero, F., Grundmann, H., Bonten, M., "Global Spread of Vancomycin-Resistant *Enterococcus faecium* from Distinct Nosocomial Genetic Complex." *Emerging Infectious Diseases* 2005; 11:821–828.

Chapter 10

189. Seppala, H., Klaukka, T., Vuopio-Varkila, J., Maotiala, A., Helenius, H., Lager, K. "The Effect of Changes in the Consumption of Macrolide Antibiotics on Erythromycin Resistance in Group A Streptococci in Finland." *New England Journal of Medicine* 1997; 337:441–446.

190. Kristinsson, K. "Effect of Antimicrobial Use and Other Risk Factors on Antimicrobial Resistance in Pneumococci." *Microbial Drug Resistance* 1997; 3:117–123.

191. Deschepper, R. VSR, Haaijer-Ruskamp, F.M. "Cross-Cultural Differences in Lay Attitudes and Utilisation of Antibiotics in a Belgian and a Dutch City." *Patient Education and Counseling* 2002; 48:161–169.

192. Davey, P., Brown, E., Fenelon, L., Finch, R., Gould, I., Holmes, A., Ramsay, C., Taylor, E., Wiffen, P., Wilcox, M. "Systematic Review of Antimicrobial Drug Prescribing in Hospitals." *Emerging Infectious Diseases* 2006; 12:211–216.

193. Harbarth, S., Albrich, W., Brun-Buisson, C. "Outpatient Antibiotic Use and Prevalence of Antibiotic-Resistant Pneumococci in France and Germany: a Sociocultural Perspective." *Emerging Infectious Diseases* 2002; 8:1460–1467.

194. Mathema, B., Cross, E., Dun, E., Park, S., Dedell, J., Slade, B., Williams, M., Riley, L., Chaturvedi, V., Perlin, D.S. "Prevalence of Vaginal Colonization by Drug-Resistant *Candida* Species in College-Age Women with Previous Exposure to Over-the-Counter Azole Antifungals." *Clinical Infectious Diseases* 2001; 33:e23–e27.

195. Phillips, I., Casewell, M., Cox, T., DeGroot, B., Friis, C., Jones, R., Nightingale, C., Preston, R., Waddell, J. "Does the Use of Antibiotics in Food Animals Pose a Risk to Human Health? A Critical Review of Published Data." *Journal of Antimicrobial Chemotherapy* 2004; 53:28–52.

196. Castanon, J. "History of the Use of Antibiotic as Growth Promoters in European Poultry Feeds." *Poultry Science* 2007; 86:2466–2471.

197. Anonymous. "Avoiding Antibiotic Resistance: Denmark's Ban on Growth Promoting Antibiotics in Food Animals." *The PEW Charitable Trusts*, 2009.

198. Baquero, F., Martinex, J.-L., Canton, R. "Antibiotics and Antibiotic Resistance in Water Environments." *Current Opinion in Biotechnology* 2008; 19:260–265.

199. Lin, A., Yu ,T-H, Lin, C-F. "Pharmaceutical Contamination in Residential, Industrial, and Agricultural Waste Streams: Risk to Aqueous Environments in Taiwan." *Chemosphere* 2008; 74:131–141.

200. Duong, H., Pham, N., Nguyen, H., Hoang, T., Pham, H., Pham, V., Berg, M., Giger, W., Alder, A. "Occurrence, Fate and Antibiotic Resistance of Fluoroquinolone Antibacterials in Hospital Wastewaters in Hanoi, Vietnam." *Chemosphere* 2008; 72:968–973.

201. Baquero, F., Martínez, J., Cantón. R. "Antibiotics and Antibiotic Resistance in Water Environments." *Current Opinion in Biotechnology* 2008; 19:260–265.

202. Verweij, P., Mellado, E., Melchers, W. "Multiple-Triazole-Resistant Aspergillosis." *New England Journal of Medicine* 2007; 356:1481–1483.

203. Enserink, M. "Farm Fungicides Linked to Resistance in a Human Pathogen." *Science* 2009; 326:1173.

204. Stratton, C. "Dead Bugs Don't Mutate: Susceptibility Issues in the Emergence of Bacterial Resistance." *Emerging Infectious Diseases* 2003; 9:10–16.

205. Lipsitch, M., Levin, B. "The Population Dynamics of Antimicrobial Chemotherapy." *Antimicrobial Agents and Chemotherapy* 1997; 41:363–373.

206. Gumbo, T., Louie, A., Deziel, M., Drusano, G.L. "Pharmacodynamic Evidence That Ciprofloxacin Failure Against Tuberculosis Is Not Due to Poor Microbial Kill but to Rapid Emergence of Resistance." *Antimicrobial Agents and Chemotherapy* 2005; 49:3178–3181.

207. Tam, V., Louie, A., Fritsche, T., Deziel, M., Liu, W., Brown, D., Deshpande, L., Leary, R., Jones, R., Drusano, GT. "Impact of Drug-Exposure intensity and Duration of Therapy on the Emergence of *Staphylococcus aureus* Resistance to a Quinolone Antimicrobial." *Journal of Infectious Diseases* 2007; 195:1818–1827.

208. Guillemot, D., Carbon, C., Balkau, B., Geslin, P., Lecoeur, H., Vauzelle-Kervroedan, F., Bouvenot, G., Eschwege, E. "Low Dosage and Long Treatment Duration of β-lactam: Risk Factors for Carriage of Penicillin-Resistant *Streptococcus pneumoniae*." *JAMA* 1998; 279:365–370.

209. Dong, Y., Zhao, X., Domagala, J., Drlica, K. "Effect of Fluoroquinolone Concentration on Selection of Resistant Mutants of *Mycobacterium bovis* BCG and *Staphylococcus aureus*." *Antimicrobial Agents and Chemotherapy* 1999; 43:1756–1758.

210. Firsov, A., Smirnova, M., Lubenko, I., Vostrov, S., Portnoy, Y., Zinner, S. "Testing the Mutant Selection Window Hypothesis with *Staphylococcus aureus* Exposed to Daptomycin and Vancomycin in an *in vitro* Dynamic Model." *Journal of Antimicrobial Chemotherapy* 2006; 58:1185–1192.

211. Hansen, G., Metzler. K., Drlica, K., Blondeau, J.M. "Mutant Prevention Concentration for Gemifloxacin with Clinical Isolates of *Streptococcus pneumoniae*." *Antimicrobial Agents and Chemotherapy* 2003; 47:440–441.

212. Jumbe, N., Louie, A., Leary, R., Liu, W., Deziel, M., Tam, V., Bachhawat, R., Freeman, C., Kahn, J., Bush, K., Dudley, M., Miller, M., Drusano, G. "Application of a Mathematical Model to Prevent in Vivo Amplification of Antibiotic-Resistant Bacterial Populations During Therapy." *Journal of Clinical Investigation* 2003; 112:275–285.

213. Drlica, K. "The Mutant Selection Window and Antimicrobial Resistance." *Journal of Antimicrobial Chemotherapy* 2003; 52:11–17.

214. Marcusson, L., Olofsson, S., Lindgren, P., Cars, O., Hughes, D. "Mutant Prevention Concentration of Ciprofloxacin for Urinary Tract Infection Isolates of *Escherichia coli*." *Journal of Antimicrobial Chemotherapy* 2005; 55:938–943.

215. Drlica, K., Zhao, X., Blondeau, J., Hesje, C. "Low Correlation Between Minimal Inhibitory Concentration (MIC) and mutant Prevention Concentration (MPC)." *Antimicrobial Agents and Chemotherapy* 2006; 50:403–404.

216. Tam, V., Louie, A., Deziel, M., Liu, W., Drusano. G. "The Relationship Between Quinolone Exposures and Resistance Amplification Is Characterized by an Inverted U: A New Paradigm for Optimizing Pharmacodynamics to Counterselect Resistance." *Antimicrobial Agents and Chemotherapy* 2007; 51:744–747.

217. Li, X., Zhao, X., Drlica, K. "Selection of *Streptococcus pneumoniae* Mutants Having Reduced Susceptibility to Levofloxacin and Moxifloxacin." *Antimicrobial Agents and Chemotherapy* 2002; 46:522–524.

218. Drlica, K., Zhao, X., Wang, J-Y, Malik, M., Lu, T., Park, S., Li, X., Perlin, D. "An Anti-Mutant Approach for Antimicrobial Use." In: Fong, I., Drlica, K., eds. *Antimicrobial Resistance and Implications for the 21st Century.* New York City: Springer, 2008:371–400.

219. Bast, D., Low, D., Duncan, C., Kilburn, L., Mandell, L., Davidson, R., de Azavedo, J. "Fluoroquinolone Resistance in Clinical Isolates of *Streptococcus pneumoniae:* Contributions of Type II Topoisomerase mutations and Efflux on Levels of Resistance." *Antimicrobial Agents and Chemotherapy* 2000; 44:3049–3054.

220. Hedlin, P., Blondeau, J. "Comparative Minimal Inhibitory and Mutant Prevention Drug Concentrations of Four Fluoroquinolones Against Ocular Isolates of *Haemophilus influenzae*." *Eye and Contact Lens* 2007; 33:161–164.

221. Fox, W., Elklard, G., Mitchison, D. "Studies on the Treatment of Tuberculosis Undertaken by the British Medical Research Council Tuberculosis Units, 1946–1986, with Relevant Subsequent Publications." *International Journal of Tuberculosis and Lung Disease* 1999; 3:S231–S279.

222. Vernon, A., Burman, W., Benator, D., Khan, A., Bozeman, L. "Acquired Rifamycin Monoresistance in Patients with HIV-Related Tuberculosis Treated with Once-Weekly Rifapentine and Isoniazid." *The Lancet* 1999; 353:1843–1847.

223. Ince, D., Zhang, X., Silver, L.C., Hooper, DC. "Dual Targeting of DNA gyrase and Topoisomerase IV: Target Interactions of Garenoxacin (BMS-284756, T-3811ME), a New Desfluoroquinolone." *Antimicrobial Agents and Chemotherapy* 2002; 46:3370–3380.

224. Fisher, L.M., Heaton, V.J. "Dual Activity of Fluoroquinolones Against *Streptococcus pneumoniae*." *Journal of Antimicrobial Chemotherapy* 2003; 51:463–464.

225. Strahilevitz, J., Hooper, D.C. "Dual Targeting of Topoisomerase IV and Gyrase to Reduce Mutant Selection: Direct Testing of the Paradigm by Using WCK-1734, a New Fluoroquinolone, and Ciprofloxacin." *Antimicrobial Agents and Chemotherapy* 2005; 49:1949–1956.

226. Robertson, G., Bonventre, E., Doyle, T., Du, Q., Duncan, L., Morris, T., Roche, E., Yan, D., Lynch, A. "In Vitro Evaluation of CBR-2092, a Novel Rifamycin-Quinolone Hybrid Antibiotic: Studies of the Mode of Action in *Staphylococcus aureus*." *Antimicrobial Agents and Chemotherapy* 2008; 52:2313–2323.

227. Robertson, G., Bonventre, E., Doyle, T., Du, Q., Duncan, L., Morris, T., Roche, E., Yan, D., Lynch, A. "In Vitro Evaluation of CBR-2092, a Novel Rifamycin-Quinolone Hybrid Antibiotic: Microbiology Profiling Studies with Staphylococci and Streptococci." *Antimicrobial Agents and Chemotherapy* 2008; 52:2324–2334.

228. Gutierrez, D. "Tyson Foods Injects Chickens with Antibiotics Before They Hatch to Claim 'Raised Without Antibiotics.'" *Naturalnews.com*, November 9, 2008.

Chapter 11

229. Barry, J., Viboud,C., Simonsen, L. "Cross-Protection between Successive Waves of the 1918–1919 Influenza Pandemic: Epidemiological Evidence from US Army Camps and from Britain." *Journal of Infectious Diseases* 2008; 198:1427–1434.

230. Brammer, L., Epperson, S., Blanton, L., Dhara, R., Wallis, T., Finelli, L., Fiore, A., Gubavera, L., Bresee, J., Klimov, A., Cox, N., Doshi, S. "Update: Influenza Activity—United States," September 28—November 29, 2008. *MMWR* 2008; 57:1329–1332.

231. Lackenby, A., Thompson, C., Democratis, J. "The Potential Impact of Neuraminidase Inhibitor Resistant Influenza." *Current Opinion in Infectious Diseases* 2008; 12:626–638.

232. Hill, A., Guralnick, R., Wilson, M., Habib, F., Janies, D. "Evolution of Drug Resistance in Multiple Distinct Lineages of H5N1 Avian Influenza Infection." *Genetics and Evolution* 2009; 9:169–178.

233. Reece, P. "Neuraminidase Inhibitor Resistance in Influenza Viruses." *Journal of Medical Virology* 2007; 79:1577–1586.

234. Hauge, S., Dudman, S., Borgen, K., Lackenby, A., Hungnes, O. "Oseltamivir-Resistant Influenza Viruses A (H1N1), Norway, 2007–08." *Emerging Infectious Diseases* 2009; 15:155–162.

235. Meijer, A., Lackenby, A., Hungnes, O., Lina, B., van der Werf, S., Schweiger, B., Opp, M., Paget, J., van de Kassteele, J., Hay, A., Zambon, M., and European Influenza Surveillance Scheme. "Oseltamivir-Resistant Influenza Virus A (H1N1), Europe, 2007–08 Season." *Emerging Infectious Diseases* 2009; 15:552–560.

236. Cheng, P., To, A., Leung, T., Leung, P., Lee, C., Lim, W. "Oseltamivir- and Amantadine-Resistant Influenza Virus A (H1N1)." *Emerging Infectious Diseases* 2010; 16:155–156.

237. Taubenberger, J., Morens, D. "1918 Influenza: the Mother of All Pandemics." *Emerging Infectious Diseases* 2006; 12:15–22.

238. Gupta, R., George, R., Nguyen van Tam, J. "Bacterial Pneumonia and Pandemic Influenza Planning." *Emerging Infectious Diseases* 2008; 14:1187–1192.

239. Shoham, D. "Review: Molecular Evolution and the Feasibility of an Avian Influenza Virus Becoming a Pandemic Strain—a Conceptual Shift." *Virus Genes* 2006; 33:127–132.

240. Vijaykrishna, D., Bahl, J., Riley, S., Duan, L., Zhang, J., Chen, H., Peiris, J., Smith, G., Guan, Y. "Evolutionary Dynamics and Emergence of Panzootic H5N1 Influenza Viruses." *PloS Pathogens* 2008; 4:e1000161.

241. Nguyen, T.D., Nguyen, T.V., Vijaykrishna, D., Webster, R., Guan, Y., Peiris, J., Smith, G. "Multiple Sublineages of Influenza A Virus (H5N1), Vietnam, 2005–2007." *Emerging Infectious Diseases* 2008; 14:632–636.

242. de Jong, M., Tran, T., Truong, H., Vo, M., Smith, G., Nguyen, V., Bach, V., Phan, T., Do, D., Guan, Y., Peiris, J., Tran, T., Farrar, J. "Oseltamivir Resistance During Treatment of Influenza A (H5N1) Infection." *New England Journal of Medicine* 2005; 353:1729–1732.

243. Brundage, J.F., Shanks, G.D. "Deaths from Bacterial Pneumonia During 1918–19 Influenza Pandemic." *Emerging Infectious Diseases* 2008; 14:1193–1199.

244. Klugman, K., Astley, C., Lipsitch, M. "Time from Illness Onset to Death, 1918 Influenza and Pneumococcal Pneumonia." *Emerging Infectious Diseases* 2009; 15:346–347.

245. Doshi, P. "Popular and Scientific Attitudes Regarding Pandemic Influenza." *Emerging Infectious Diseases* 2008; 14:1501–1502.

246. Ortiz, J.R., Kamimoto, L., Aubert, R., Yao, J., Shay, D., Bresee, J., Epstein, R., "Oseltamivir Prescribing in Pharmacy-Benefits Database, United States, 2004–2005." *Emerging Infectious Diseases* 2008; 14:1280–1283.

247. Green, M., Nettey, H., Wirtz, R. "Determination of Oseltamivir Quality by Colorimetric and Liquid Chromatographic Methods." *Emerging Infectious Diseases* 2008; 14:552–556.

Chapter 12

248. Sahm, D., Karlowsky, J., Kelly, L., Critchley, I., Jones, M., Thornsberry, C., Mauriz, Y., Kahn, J. "Need for Annual Surveillance of Antimicrobial Resistance in *Streptococcus pneumoniae* in the United States: 2-Year Longitudinal Analysis." *Antimicrobial Agents and Chemotherapy* 2001; 45:1037–1042.

249. Karlowsky, J., Thornsberry, C., Jones, M., Evangelista, A., Critchley, I., Sahm, D. "Factors Associated with Relative Rates of Antimicrobial Resistance Among *Streptococcus pneumoniae* in the United States: Results from the TRUST Surveillance Program (1998–2002)." *Clinical Infectious Diseases* 2003; 36.

250. Valentine, V. "A Timeline of Andrew Speaker's Infection." www.npr.org./news/specials/tb/, June 6, 2007.

251. van der Sande, M., Teunis, P., Sabel, R. "Professional and Home-Made Face Masks Reduce Exposure to Respiratory Infections Among the General Population." *PLoS ONE* 2008; 3:e2618.

252. Casanova, L., Alfano-Sobsey, E., Rutala, W., Weber, D., Sobsey, M. "Virus Transfer from Personal Protective Equipment to Healthcare Employees' Skin and Clothing." *Emerging Infectious Diseases* 2008; 14:1291–1293.

253. Casanova, L., Alfano-Sobsey, E., Rutala, W., Weber, D., Sobsey, M. "Virus Transfer from Personal Protective Equipment to Healthcare Employees' Skin and Clothing." *Emerging Infectious Diseases* 2008; 14:1291–1293.

254. Bayard, V., Kitsutani, P., Barria, E., Ruedas, L., Tinnin, D., Muñoz, C., de Mosca, I., Guerrero, G., Kant, R., Garcia, A., Caceres, L., Gracio, F., Quiroz, E., de Castillo, Z., Armien, B., Libel, M., Mills, J. "Outbreak of Hantavirus Pulmonary Syndrome, Los Santos, Panama, 1999–2000." *Emerging Infectious Diseases* 2004; 10:1635–1642.

255. Martinez, V., Bellomo, C., San-Juan, J., Pinna, D., Forlenza, R., Elder, M., Padula, P. "Person-to-Person Transmission of Andes Virus." *Emerging Infectious Diseases* 2005; 11:1848–1853.

256. Zhang, Y., Dong, X., Li, X., Ma, C., Xiong, H., Yan, G., Gao, N., Jiang, D., Li, M., Li, L., Zou, Y., Plyusnin, A. "Seoul Virus and Hantavirus Disease, Shenyang, People's Republic of China." *Emerging Infectious Diseases* 2009; 15:200–206.

257. Bearden, D., Allen, G., Christensen, J. "Comparative in Vitro Activities of Topical Wound Care Products Against Community-Associated Methicillin-Resistant *Staphylococcus aureus*." *Journal of Antimicrobial Chemotherapy* 2008; 62:769–772.

258. Huijsdens, X., Janssen, M., Renders, N., Leenders, A., van Wijk, P., van Santen-Verheuvel, M., van Driel, J., Morroy, G. "Methicillin-Resistant *Staphylococcus aureus* in a Beauty Salon, the Netherlands." *Emerging Infectious Diseases* 2008; 14:1797–1799.

259. Weese, J., Archambault, M., Willey, B., Hearn, P., Kreiswirth, B., Said-Salim, B., McGeer, A., Likhoshvay, Y., Prescott, J., Low, D. "Methicillin-Resistant *Staphylococcus aureus* in Horses and Horse Personnel, 2000–2002." *Emerging Infectious Diseases* 2005; 11:430–435.

260. Baptiste, K., Williams, K., Willams, N., Wattret, A., Clegg, P., Dawson, S., Corkill, J., O'Neill, T., Hart, C. "Methicillin-Resistant Staphylococci in Companion Animals." *Emerging Infectious Diseases* 2005; 11:1942–1944.

261. Lewis, H.C., Mølbak, K., Reese, C., Aarestrup, F., Selchau, M., Sørum, M., Skov, R. "Pigs as Source of Methicillin-Resistant *Staphylococcus aureus* CC398 Infections in Humans, Denmark." *Emerging Infectious Diseases* 2008; 14:1383–1389.

262. van Loo, I., Diederen, B., Savelkoul, P., Woudenberg, J., Roosendaal, R., van Belkum, A., Lemmens den Toom, N., Verhulst, C., van Keulen, P., Kluytmans, J. "Methicillin-Resistant *Staphylococcus aureus* in meat Products, the Netherlands." *Emerging Infectious Diseases* 2007; 13:1753–1755.

263. Persoons, D., VanHoorebeke, S., Hermans, K., Butaye, P., de Kruif, A., Haesebrouck, F., Dewulf, J. "Methicillin-Resistant *Staphylococcus aureus* in Poultry." *Emerging Infectious Diseases* 2009; 15:452–453.

264. David, M., Mennella, C., Mansour, M., Boyle-Vavra, S., Daum, R. "Predominance of Methicillin-Resistant *Staphylococcus aureus* Among Pathogens Causing Skin and Soft Tissue Infections in a Large Urban Jail: Risk Factors and Recurrence Rates." *Journal of Clinical Microbiology* 2008; 46:3222–3227.

265. Diep, B., Chambers, H., Graber, C., Szumowski, J., Miller, L., Han, L., Chen, J., Lin, F., Lin, J., Phan, T., Carleton, H., McDougal, L., Tenover, F., Cohen, D., Mayer, K., Sensabaugh, G., Perdreau-Remington, F. "Emergence of Multidrug-Resistant, Community-Associated, Methicillin-Resistant *Staphylococcus aureus* Clone USA300 in Men Who Have Sex with Men." *Annals of Internal Medicine* 2008; 148:249–257.

266. Matsumoto, T., Muratani, T., Takahashi, K., Ikuyama, T., Yokoo, D., Ando, Y., Sato, Y., Kurashima, M., Shimokawa, H., Yanai, S. "Multiple Doses of Cefodizime Are Necessary for the Treatment of *Neisseria gonorrhoeae* Pharyngeal Infection." *Journal of Infection and Chemotherapy* 2008; 12:145–147.

267. Workowski, K., Berman, S., Douglas, J. "Emerging Antimicrobial Resistance in *Neisseria gonorrhoeae:* Urgent Need to Strengthen Prevention Strategies. *Annals of Internal Medicine* 2008; 149:363–364.

268. Matsumoto, T., Muratani, T., Takahashi, K., Ikuyama, T., Yokoo, D., Ando, Y., Sato, Y., Kurashima, M., Shimokawa, H., Yanai, S. "Multiple Doses of Cefodizime Are Necessary for the Treatment of *Neisseria gonorrhoeae* Pharyngeal Infection." *Journal of Infection and Chemotherapy* 2006; 12:145–147.

269. Allen, G., Hankins, C. "Evaluation of the Mutant Selection Window for Fluoroquinolones Against *Neisseria gonorrhoeae*." *Journal of Antimicrobial Chemotherapy* 2009; 64:359–363.

270. Stäger, K., Legros, F., Krause, G., Low, N., Bradley, D., Desai, M., Graf, S., D'Amato, S., Mizuno, Y., Janzon, R., Petersen, E., Kester, J., Steffen, R., Schlagenhauf, P. "Imported Malaria in Children in Industrialized Countries, 1992–2002." *Emerging Infectious Diseases* 2009; 15:185–191.

271. Askling, H., Nilsson, J., Tegnell, A., Janzon, R., Ekdahl, K. "Malaria Risk in Travelers." *Emerging Infectious Diseases* 2005; 11:436–441.

272. Mason, M. "Quality of Malaria Pills in Africa Raises Resistance Fears." *Miami Herald Tribune* International Edition. Miami, FL, February 9, 2010:8A.

273. Mumcuoglu, K., Hemingway, J., Miller, J., Ioffe-Uspensky, I., Klaus, S., Ben-Ishai, F., Galun, R. "Permethrin Resistance in the Head Louse Pediculus Capitis from Israel." *Medical and Veterinary Entomology* 1995; 9:427–432.

274. Adak, G., Meakins, S., Yip, H., Lopman, B.A., O'Brien, S.J. "Disease Risks from Foods, England and Wales, 1996–2000." *Emerging Infectious Diseases* 2005; 11:365–372.

275. Mead, P., Slutsker, L., Dietz, V., McCaig, L., Bresee, J., Shapiro, C., Griffin, P., Tauxe, R. "Food-Related Illness and Death in the United States." *Emerging Infectious Diseases* 1999; 5:607–625.

276. Manges, A., Johnson, J., Foxman, B., O'Bryan, T., Fullerton, K., Riley, L. "Widespread Distribution of Urinary Tract Infections Caused by a Multidrug-Resistant *Escherichia coli* Clonal Group." *New England Journal of Medicine* 2001; 345:1007–1013.

277. Johnson, J., McCabe, J., White, D., Johnston, B., Kuskowski, M., McDermott, P. "Molecular Analysis of *Escherichia coli* from Retail Meats (2002–2004) from the United States National Antimicrobial Resistance Monitoring System." *Clinical Infectious Diseases* 2009; 49:195–201.

278. Rabatsky-Ehr, T., Whichard, J., Rossiter, S., Holland, B., Stamey, K., Headrick, M., Barrett, T., Angulo, F. and the NARMS Working Group. "Multidrug-Resistant Strains of *Salmonella enterica* Typhimurium, United States, 1997–1998." *Emerging Infectious Diseases* 2004; 10:795–801.

279. Ryan, C., Nickels, M., Hargrett-Bean, N., Potter, M., Endo, T., Mayer, L., Langkop, C., Gibson, C., McDonald, R., Kenney, R. "Massive Outbreak of Antimicrobial-Resistant Salmonellosis Traced to Pasteurized Milk." *JAMA* 1987; 258:3269–3274.

280. Olsen, S., Ying, M., Davis M., Deasy, M., Holland, B., Iampietro, L., Baysinger, C., Sassano, F., Polk, L., Gormley, B., Hung, M., Pilot, K., Orsini, M., Van Duyne, S., Rankin, S., Genese, C., Bresnitz, E., Smucker, J., Moll, M., Sobel J. "Multidrug-Resistant *Salmonella typhimurium* Infection from Milk Contaminated after Pasteurization." *Emerging Infectious Diseases* 2004; 10:932–935.

281. Wedel, S., Bender, J., Leano, F., Boxrud, D., Hedberg, C., Smith, K. "Antimicrobial-Drug Susceptibility of Human and Animal *Salmonella typhimurium*, Minnesota, 1997–2003." *Emerging Infectious Diseases* 2005; 11:1899–1906.

282. Fey, P., Safranek, T., Rupp, M., Dunne, E., Ribot, E., Iwen, P., Bradford, P., Angulo, F., Hinrichs, S. "Ceftriaxone-Resistant Salmonella Infection Acquired by a Child from Cattle." *New England Journal of Medicine* 2000; 342:1242–1249.

283. Helms, M., Vastrup, P., Gerner-Smidt, P., Mølbak, K. "Excess Mortality Associated with Antimicrobial Drug-Resistant *Salmonella typhimurium*." *Emerging Infectious Diseases* 2002; 8:490–495.

284. Molbak, K. "Human Health Consequences of Antimicrobial Drug-Resistant *Salmonella* and Other Foodborne Pathogens." *Clinical Infectious Diseases* 2005; 41:1613–1620.

285. Nichols, G. "Fly Transmission of *Campylobacter*." *Emerging Infectious Diseases* 2005; 11:361–364.

286. Hald, B., Sommer, H., Skovgård, H. "Use of Fly Screens to Reduce *Campylobacter* spp. Introduction in Broiler Houses." *Emerging Infectious Diseases* 2007; 13:1951–1953.

287. Keen, J.E., Wittum, T., Dunn, J., Bono, J., Durso, L. "Shiga-Toxigenic *Escherichia coli* O157 in Agricultural Fair Livestock, United States." *Emerging Infectious Diseases* 2006; 12:780–786.

288. Hsueh, P., Teng, L., Tseng, S, Chang, C., Wan, J., Yan, J., Lee, C., Chuang, Y., Huang, W., Yang, D., Shyr, J., Yu, K., Wang, L., Lu, J., Ko, W., Wu, J., Chang, F. "Ciprofloxacin-Resistant *Salmonella enterica* Typhimurium and Choleraesuis from Pigs to Humans, Taiwan." *Emerging Infectious Diseases* 2004; 10:60–68.

289. Hald, B., Skovgård, H., Bang, D., Pedersen, K., Dybdahl, J., Jespersen, J., Madsen, M. "Flies and *Campylobacter* Infection of Broiler Flocks." *Emerging Infectious Diseases* 2004; 10:1490–1492.

290. Riedner, G., Rusizoka, M., Todd, J., Maboko, L., Hoelscher, M., Mmbando, E., Samky, E., Lyamuya, E., Mabey, D., Grosskurth, H., Hayes, R. "Single-Dose Azithromycin Versus Penicillin G Benzathine for the Treatment of Early Syphilis." *New England Journal of Medicine* 2005; 353:1236–44.

291. Katz, K., Klausner, J. "Azithromycin Resistance in *Treponema pallidum*." *Current Opinion in Infectious Diseases* 2008; 21:83–91.

292. Walker, P., Reynolds, M., Ashbee, H., Brown, C., Evans, E. "Vaginal Yeasts in the Era of 'Over the Counter' Antifungals." *Sexually Transmitted Infections* 2000; 76:437–438.

293. Sieradzki, K., Roberts, R., Haber, S., Tomasz, A. "The Development of Vancomycin Resistance in a Patient with Methicillin Resistant *Staphylococcus aureus* Infection." *New England Journal of Medicine* 1999; 340:517–523.

294. Aiello, A., Larson, E., Levy, S. "Consumer Antibacterial Soaps: Effective or Just Risky?" *Clinical Infectious Diseases* 2007; Suppl 2:S137–S147.

295. Shehab, N., Patel, P., Srinivasan, A., Budnitz, D. "Emergency Department Visits for Antibiotic-Associated Adverse Events." *Clinical Infectious Diseases* 2008; 47:735–743.

296. Modena, S., Gollamudi, S., Friedenberg, F. "Continuation of Antibiotics is Associated with Failure of Metronidazole for *Clostridium difficile*-Associated Diarrhea." *Journal of Clinical Gastroenterology* 2006; 40:49–54.

297. Ho, P.L., Tse, W.S., Tsang, K., Kwok, T., Ng, T., Cheng, V., Chan, R. "Risk Factors for Acquisition of Levofloxacin-Resistant *Streptococcus pneumoniae*: a Case-Control Study." *Clinical Infectious Diseases* 2001; 32:701–707.

Afterword

298. Hammerum, A., Heuer, O., Emborg, H., Bagger-Skjøt, L., Jensen, V., Rogues, A., Skov, R., Agersø, Y., Brandt, C., Seyfarth, A., Muller, A., Hovgaard, K., Ajufo, J., Bager, F., Aarestrup, F., Frimodt-Møller, N., Wegener, H. "Danish Integrated Antimicrobial Resistance Monitoring and Research Program." *Emerging Infectious Diseases* 2007; 13:1632–1639.

299. Lyte, M., Freestone, P. "Microbial Endocrinology Comes of Age." Microbe 2009; 4:169–175.

Appendix B

300. Chang, D., Anderson, S., Wannemuehler, K., Engelthaler, D., Erhart, L., Sunenshine, R., Burwell, L., Park, B. "Testing for Coccidioidomycosis Among Patients with Community-Acquired Pneumonia." *Emerging Infectious Diseases* 2008; 14:1053–1059.

301. Gauthier, G., Klein, B.S. "Insights into Fungal Morphogenesis and Immune Evasion." *Microbe* 2008; 3:416–423.

Index

B

C

D

E

N

Q

R

T

U

V

FT Press
FINANCIAL TIMES